APPLICATIONS OF ATOMIC SPECTROMETRY TO REGULATORY COMPLIANCE MONITORING

APPLICATIONS OF ATOMIC SPECTROMETRY TO REGULATORY COMPLIANCE MONITORING

Second Edition

STEPHEN W. JENNISS
SIDNEY A. KATZ
RICHARD W. LYNCH

New York • Chichester • Weinheim • Brisbane • Singapore • Toronto

Stephen W. Jenniss
Director
State of New Jersey Department of Health
Environmental and Chemical Laboratory Services
Trenton, NJ 08625-0360

Sidney A. Katz
Department of Chemistry
Rutgers University
Camden, NJ 08102

Richard W. Lynch
Director
21st Century Environmental
618 Heron Drive
Bridgeport, NJ 08014

This book is printed on acid-free paper.

Library of Congress Cataloging in Publication Data:

Jenniss, Stephen W., 1952–
Applications of atomic spectrometry to regulatory compliance monitoring / Stephen W. Jenniss, Sidney A. Katz, Richard W. Lynch. — 2nd ed.
p. cm.
Second ed. of: Regulatory compliance monitoring by atomic absorption spectroscopy. c 1983.
Includes index.
ISBN 0-471-19039-X (cloth : alk. paper)
1. Atomic absorption spectroscopy. 2. Environmental monitoring. I. Katz, Sidney, A., 1935– . II. Lynch, Richard W. III. Katz, Sidney A., 1935– . Regulatory compliance monitoring by atomic absorption spectroscopy. IV. Title.
QD96.A8K37 1997
628.5′2—dc21

96-51619
CIP

Printed in the United States of America

ISBN 0-471-19039-X Wiley-VCH, Inc.

10 9 8 7 6 5 4 3 2 1

CONTENTS

PREFACE

Human exposure to and intake of trace metals are areas of high activity from the toxicological as well as the nutritional perspectives. State and federal public health and environmental protection agencies require or at least recommend monitoring the occupational, community, and domestic environments for toxic trace metals. Some federal statutes requiring compliance monitoring are the Occupational Health and Safety Act of 1970 (PL 91-596), the Clean Air Act of 1970, the Safe Drinking Water Act of 1972 (PL 93-523), the Resource Conservation and Recovery Act of 1976 (PL-94-580), the Clean Water Act of 1977 (PL 95-217), the Water Quality Act of 1987 (PL 100-4), and the Nutritional Labeling and Education Act.

Applications of Atomic Spectrometry to Regulatory Compliance Monitoring is a sequel to *Regulatory Compliance Monitoring by Atomic Absorption Spectroscopy* (Verlag Chemie International, 1983). While atomic absorption spectrometry (AAS) remains a popular technique for quantifying the many metallic constituents of environmental samples, inductively coupled plasma–atomic emission spectrometry (ICP-AES) and inductively coupled plasma–mass spectrometry (ICP-MS) have been adopted and are now approved for this purpose. The sequel is necessitated by this and other significant changes made during the last decade in the procedures required and/or recommended by regulatory agencies in the United States for the collection, preservation, storage, and preparation of samples and for the quantification of the metallic contaminants in these samples. Such regulatory bodies include the Environmental Protection Agency, the Food and Drug Administration, and the Occupational Safety and Health Administration.

During the last decade, the regulated community has grown to include, among others, food processors, who must now monitor sodium in their consumer products. This growth, coupled with significant changes in the regulations, is an indicator that the sequel will become an important complement to the original volume, *Regulatory Compliance Monitoring by Atomic Absorption Spectroscopy.*

One of the objectives of *Applications of Atomic Spectrometry to Regulatory Compliance Monitoring* is to put into the hands of the practitioner a single book that will allow him or her to generate scientifically and legally defensible laboratory results for regulatory compliance monitoring of metallic contaminants.

The preparation of *Regulatory Compliance Monitoring by Atomic Absorption Spectroscopy* could not have been accomplished without the support and encouragement of Shelly Jenniss, Sheila Katz, and Pat Lynch. We are grateful for their patience and endurance while this work was under way.

STEPHEN W. JENNISS
SIDNEY A. KATZ
RICHARD W. LYNCH

1

INTRODUCTION

Continued research and development in environmental analytical chemistry is responsible for several major advances in the applications of atomic spectrometry to regulatory compliance monitoring. Microwave-assisted sample preparation and flow injection analysis have become important adjuncts to the determination of metallic contaminants by conventional flame and furnace atomic absorption spectrometry. The hydride and cold vapor atomic absorption techniques have also benefited from the adoption of these enhancements. The adoption of inductively coupled plasma–atomic emission spectrometry by regulatory agencies has introduced simultaneous multielement determinations, and the inclusion of inductively coupled plasma–mass spectrometry in the regulations has made possible the direct, simultaneous determination of metallic contaminants below the parts-per-billion level. Methods 3015 and 3051 of the United States Department of Environmental Protection (US EPA) require microwave-assisted acid digestion for the preparation of solid and liquid samples prior to determination of metal contents by atomic spectrometry. US EPA Methods 200.8 and 200.10 are among those with which the concentrations of many elements in aqueous samples may be determined by inductively coupled plasma–mass spectrometry. Grosser[1] has described some of these developments in her overview on the advances in inorganic laboratory instrumentation for US EPA methods. Riley et al.[2] have described the development by the federal Department of Energy (US DOE) of analytical methods document for the determination of ^{99}Tc, ^{230}Th, and ^{234}U and other radioactive nuclides utilizing inductively coupled plasma–mass spectrometry with flow injection preconcentration. These are only some of the technological advances in regulatory compliance monitoring from the recent past. Regulatory compliance monitoring is and will continue to be viable and dynamic with continual changes to allow the use of the best available technology.

1.1 ATOMIC ABSORPTION SPECTROMETRY

Atomic absorption spectrometry has stood the test of time for the determination of trace elements, usually metals, in a wide variety of environmental and biological matrices. The success of atomic absorption spectrometry was due not only to its sensitivity and selectivity, but also to its speed, simplicity, and broad scope.

1.1.1. Scope and Sensitivity of Atomic Absorption Spectrometry

More than six dozen elements can be determined by atomic absorption spectrometry. In many cases, the detection limits range from a few tenths of a part per billion (when electrothermal atomization techniques are employed) to a few tenths of a part per million (using conventional flame atomization). Detection limits with electrothermal and/or gaseous hydride atomization techniques are often three orders of magnitude lower than those established with conventional flame atomization. The experimental conditions and optimum ranges for some two dozen elements of biological and environmental significance are summarized in Table 1.1, and the regulatory limits for some trace elements in air and water are listed in Table 1.2. The detection limits for atomic absorption spectrometry are compared to those of the other atomic spectrometric techniques in Table 1.3.

1.1.2 Theory of Atomic Absorption Spectrometry

Atomic absorption spectrometry is based upon the absorption of resonance radiation by an atomic vapor of the analyte. The resonance radiation corresponds to the wavelengths associated with the excitation of ground state gaseous analyte atoms. Most elements show relatively simple, well-defined, characteristic absorption spectra. There is little likelihood of spectral interference when the excitation radiation is of high spectral purity. Hence, atomic absorption is highly selective.

The extent to which an atomic vapor of thickness ℓ absorbs resonance radiation of frequency ν is:

$$\frac{I_\nu}{I_0} = e^{-k\nu\ell}$$

and

$$\int_0^\infty k_\nu d\nu = \left(\frac{\pi e^2}{mc}\right) N_0 f$$

where k is the absorption coefficient of the atom at frequency ν, e is the electronic charge, m is the electronic mass, N_0 is the number of analyte atoms in the ground state, and f is the oscillator strength of the absorption line. The oscillator strength f reflects the probability of a transition from the ground state to an excited state: that is,

$$f = \frac{m\,c}{8\pi^2 e^2} \frac{g_1}{g_2} \frac{A^2}{\nu}$$

where g_1 and g_2 are the statistical weights for the atoms in the ground state and the excited state, A is the Einstein coefficient, and ν is the frequency at which the transition takes place. The Boltzman distribution,

$$\frac{N_1}{N_2} = \left(\frac{g_1}{g_2}\right) e^{-\Delta E/kT}$$

in which N_1 and N_2 are the numbers of species in energy states E_1 and E_2 separated by ΔE, g_1 and g_2 are the statistical weights for the species in energy states E_1 and E_2, and k is the Boltzmann constant = 1.38×10^{-23} J/K, governs the number of atoms in each state, and, consequently, the intensity of the absorption.

When the resonance radiation characteristic of a given element is incident upon a cell containing that element as an atomic vapor, absorption occurs in the direction of the incident radiation. The excited atoms can and do return to the ground state, but the resulting emissions do not compensate for the absorption because the former is isotropic.

In a cylindrical cell of radius r and length ℓ containing C gaseous atoms per unit volume with N_0 resonance radiation photons entering per unit time per unit area, the rate of photon absorption is

$$-dN = \frac{NkC\pi r^2 dx}{\pi r^2}$$

where k is the reaction cross section for resonance absorption, and $C\pi r^2 dx$ is the number of gaseous atoms in the element of volume under consideration. A relationship between of the Bouguer–Lambert–Beer–Bernard type absorption and concentration can be obtained by integration:

$$-\int_{N_0}^{N} \left(\frac{dN}{N}\right) = kC \int_0^{\ell} dx \qquad \text{and} \qquad \ln\left(\frac{N_0}{N}\right) = kC\ell$$

The logarithm of the ratio of the incident photon flux to the emergent photon flux is the absorbance of the resonance radiation. Because the cell volume is constant, the absorbance is directly proportional to the number of ground state gaseous atoms in the system. This proportionality is the basis for quantification by measurement of atomic absorption. Conditions must be carefully controlled to ensure that this proportionality is maintained. Atomic absorption spectrometry is, obviously, a comparative rather than an absolute measurement technique.

TABLE 1.1 Parameters Associated with the Determination of Some Frequently Encountered Elements by Atomic Absorption Spectrometry

	Flame Technique				
Element	Wave-length (nm)	Fuel/ Oxidant	Flame Conditions	Modifier	Working Range (ppm)
Ag	328.1	C_2H_2/air	Lean blue		0.1–4
Al	309.2	C_2H_2/N_2O	Rich red	KCl	5–50
As	193.7	H_2/air		Hydride	0.002 –0.02
Ba	553.6	C_2H_2/air	Rich red	KCl	1–20
Be	234.9	C_2H_2/N_2O	Rich red		0.1–2
Ca	422.7	C_2H_2/air	Rich yellow	$LaCl_3$	0.1–10
Cd	228.8	C_2H_2/air	Lean blue		0.1–2
Co	204.7	C_2H_2/air	Stoichiometric		0.5–10
Cr	357.9	C_2H_2/N_2O	Rich red		0.5–10
Cu	324.7	C_2H_2/air	Lean blue		0.1–10
Fe	248.3	C_2H_2/air	Lean blue		0.3–10
Hg	253.7			Cold vapor	0.002–0.02
K	766.5	C_2H_2/air	Slightly lean		0.1–2
Mg	285.2	C_2H_2/air	Rich yellow	$LaCl_3$	0.1–2
Mn	279.5	C_2H_2/air	Lean blue		0.1–10
Mo	313.3	C_2H_2/N_2O	Rich red	$Al(NO_3)_3$	5–20
Na	589.6	C_2H_2/air	Lean blue		0.1–1
Ni	232.0	C_2H_2/air	Lean blue		0.5–10
Pb	283.3	C_2H_2/air	Slightly lean		1–20
Sb	217.6	C_2H_2/air	Lean blue		1–40
Se	196.0	H_2/air		Hydride	0.002–0.02
Sn	286.3	C_2H_2/air	Rich yellow		10–200
Ti	365.3	C_2H_2/N_2O	Rich red	KCl	5–100
Tl	276.8	C_2H_2/air	Lean blue		1–20
V	318.4	C_2H_2/N_2O	Rich red	$Al(NO_3)_3$	1–100
Zn	213.9	C_2H_2/air	Lean blue		0.1–2

1.1.3 Requirements for Atomic Absorption Spectrometry

The measurement of atomic absorbance requires a source of resonance radiation, a monochromator for isolating the resonance lines, a detector for determining the intensities of the incident and emergent photon fluxes, and an atomizer for generating the atomic vapor of analyte.

1.1.3.1 Sources of Resonance Radiation Atomic spectral lines typically have natural widths on the order of 10^{-5} nm. Isolation of light with this degree of spectral purity from a continuum source demands a monochromator of very high resolution and slit widths narrow enough to ensure minimal stimulation of the photon detector by the feeble emergent beam. The hollow cathode lamp is a simple alternative to meeting these requirements.

TABLE 1.1 *Continued*

	Furnace Technique			
	Drying times	Ashing	Atomizing	Working
Element	(s)	temperatures (°C)		Range (ppb)
Ag	30/125	30/400	10/2700	1–25
Al	30/125	30/1300	10/2700	20–200
As	30/125	30/1100	10/2700	5–100
Ba	30/125	30/1200	10/2800	10–200
Be	30/125	30/1000	10/2800	1–30
Ca				
Cd	30/125	30/500	10/1900	0.5–10
Co	30/125	30/900	10/2700	5–100
Cr	30/125	30/1000	10/2700	5–100
Cu	30/125	30/900	10/2700	5–100
Fe	30/125	30/1000	10/2700	5–100
Hg				
K				
Mg				
Mn	30/125	30/1000	10/2700	1–30
Mo	30/125	30/1400	10/2800	3–60
Na				
Ni	30/125	30/900	10/2800	5–100
Pb	30/125	30/500	10/2700	5–100
Sb	30/125	30/800	10/2700	20–300
Se	30/125	30/1200	10/2700	5–100
Sn	30/125	30/600	10/2700	20–300
Ti	30/125	30/1400	15/2800	50–500
Tl	30/125	30/400	10/2400	5–100
V	30/125	30/1400	10/2800	10–200
Zn	30/125	30/400	10/2500	0.2–4

Hollow Cathode Lamps The hollow cathode lamp consists of a glass cylinder with an appropriate optical window containing a cup-shaped cathode fabricated from an element identical to the analyte. The cylinder is filled with neon or argon at 5–10 torr. At potentials of a few hundred volts and currents of 10–20 mA, the atoms of the filling gas undergo ionization and bring about the sputtering and excitation of the cathode material within the cup. The emissions are essentially the line spectrum of the material from which it was fabricated and of the filling gas. The resonance absorption line of the analyte is contained in the emission spectrum of the cathode material.

The control of lamp current is important. As the current is increased, both sputtering and excitation of the cathode material increase. The former gives rise to a high concentration of gaseous atoms from the cathode material within the lamp.

TABLE 1.2 Regulatory Limits for Trace Elements in Air and Water

Element		Limits	
		Workplace, Airborne (mg/m³)[a]	Drinking Water (mg/L)[b]
Aluminum,	total	15	
Arsenic,	total		0.05
	Inorganic	10	
	Organic	0.5	
Barium,	total	0.5	1
Beryllium,	total	0.002	
Cadmium,	total		0.01
	Dust		0.2
	Fume	0.1	
Calcium,	total	5	
Chromium,	total		0.1
	Hexavalent	0.1	
	Trivalent	0.5	
Cobalt,	total	0.1	
Copper,	total		1
	Dust	1	
	Fume	0.1	
Iron,	total	10	0.3
Lead,	total	0.05	0.05
Magnesium,	total	15	
Manganese,	total	5	0.05
Mercury,	total	0.1	0.002
Nickel,	total	1	
Selenium,	total	0.2	0.01
Silver,	total	0.01	0.05
Thallium,	total	0.1	
Titanium,	total	15	
Vanadium,	Dust	0.5	
	Fume	0.1	
Zinc,	total		5
	Dust	15	
	Fume	5	

[a]*NIOSH Manual of Analytical Methods,* 4th ed. (Cincinnati, OH, 1994), Method 7300, p. 6.
[b]*Is Your Drinking Water Safe?* EPA5709-89-005, United States Environmental Protection Agency, Washington, DC, 1989.

These gaseous atoms absorb the resonance radiation. Self-absorption broadening results in degradation of the sensitivity and curvature of the atomic absorbance–concentration calibration line. Ideally, the current should be sufficient to produce an intense resonance line with minimal self-absorption broadening. Best guidance on lamp currents is obtained from manufacturers' literature. Special attention must be given to differentiate between the current recommendations for mechani-

TABLE 1.3 Atomic Spectrometry Detection Limits[a]

	Limits (μg/L)[b]				
Element	Flame AAS	Cold Vapor or Hydride	ET AAS	ICP-AES	ICP-MS
Ag	105		0.05	1.5	0.003
Al	45		0.3	6	0.006
As	150	0.03	0.5	30	0.006
Au	9		0.4	6	0.001
B	1000		45	3	0.09
Ba	15		0.9	0.15	0.002
Be	1.5		0.02	0.09	0.03
Bi	30	0.03	0.6	30	0.0005
Br					0.2
C				75	150
Ca	1.5		0.03	0.15	2
Cd	0.8		0.02	1.5	0.003
Ce				15	0.0004
Cl					10
Co	9		0.4	3	0.0009
Cr	3		0.08	3	0.02
Cs	15				0.0005
Cu	1.5		0.25	1.5	0.003
Dy	50				0.001
Er	60				0.0008
Eu	30				0.0007
F					1000
Fe	5		0.3	1.5	0.4
Ga	75			15	0.001
Gd	1800				0.002
Ge	300			15	0.003
Hf	300				0.0006
Hg	300	0.009	15	30	0.004
Ho	60				<0.0005
I					0.008
In	30			45	<0.0005
Ir	900		7	30	0.0006
K	3		0.02	75	1
La	3000			1.5	0.0005
Li	0.8		0.15	1.5	0.03
Lu	1000				<0.0005
Mg	0.15		0.01	0.15	0.007
Mn	1.5		0.09	0.6	0.002
Mo	45		0.2	7.5	0.003
Na	0.3		0.05	6	0.05
Nb	1500			5	0.0009
Nd	1500				0.002

(continued)

TABLE 1.3 *Continued*

	Limits (μg/L)[b]				
Element	Flame AAS	Cold Vapor or Hydride	ET AAS	ICP-AES	ICP-MS
Ni	6		0.8	6	0.005
Os	120				
P	7500		320	45	0.3
Pb	15		0.15	30	0.001
Pd	30		2	1.5	0.003
Pr	7500				<0.0005
Pt	60		5	30	0.002
Rb	3		0.08		0.003
Re	750			30	0.0006
Rh	6			30	0.0008
Ru	100		3	6	0.002
S				75	70
Sb	45	0.15	0.4	90	0.001
Sc	30			0.3	0.02
Se	100	0.03	0.7	90	0.06
Si	90		2.5	5	0.7
Sm	3000				0.001
Sn	150		0.5	60	0.002
Sr	3		0.06	0.075	0.0008
Ta	1500			30	0.0006
Tb	900				<0.0005
Te	30	0.03	1	75	0.01
Th					<0.0005
Ti	75		0.9	0.75	0.006
Tl	15		0.4	60	0.0005
Tm	15				<0.0005
U	15000			15	<0.0005
V	60		0.3	3	0.002
W	1500			30	0.001
Y	75			0.3	0.0009
Yb	8				0.001
Zn	1.5		0.3	1.5	0.003
Zr	450			1.5	0.004

[a]These detection limits were established with aqueous solutions of elemental standards using commercial instrumentation from the Perkin-Elmer Corporation (the model 5100 atomic absorption spectrometer, the FIAS-200 flow injection amalgamation system, and the MHS-10 mercury/hydride system, the model 5100 PC atomic absorption spectrometer with 5100 ZL Zeeman furnace module or the model 4100 ZL atomic absorption spectrometer under STPF conditions, the Plasma 2000 inductively coupled plasma–atomic emission spectrometer, and the ELAN 5000 inductively coupled plasma–mass spectrometer). All detection limits are based on a 98% (3 σ) confidence level.

[b]AAS, atomic absorption spectrometry; ET-AAS, electrothermal AAS; ICP-AES, inductively coupled plasma–atomic emission spectrometry; ICP-MS, ICP mass spectrometry.

Source: The Guide to Techniques and Applications of Atomic Spectrometry, Perkin-Elmer Corp., Norwalk, CT, 1993.

cally modulated spectrophotometers and those for electronically modulated instruments.

Electrodeless Discharge Lamps The electrodeless discharge lamp overcomes some of the difficulties encountered with hollow cathode lamps for the more volatile elements with resonance lines in the far-ultraviolet region of the spectrum. Electrodeless discharge lamps for arsenic and selenium produce much more intense radiation than do the corresponding hollow cathode lamps. A hollow cathode lamp operated at the current required to produce resonance radiation equal to that of an electrodeless discharge lamp has a useful life that is quite short. Electrodeless discharge lamps are not limited to arsenic and selenium. They are available for antimony, arsenic, bismuth, cadmium, cesium, germanium, lead, mercury, phosphorus, rubidium, selenium, tellurium, thallium, tin, and zinc. The operation of these lamps requires a special power supply.

Electrodeless discharge lamps produce resonance radiation by high frequency excitation of the desired element. The excitation takes place in a small, sealed quartz tube (1 cm × 5 cm) or sphere (1.5 cm diameter) containing a few milligrams of the element or one of its compounds, usually the iodide. The quartz container also contains argon, xenon, or krypton at a few torr. The quartz container is situated within the coil of a high frequency generator.

The electrodeless discharge lamp requires some 30 minutes for stabilization. After this warm-up period, the spectral output is some 10–100 times greater than that of a corresponding hollow cathode lamp. The emitted radiations are not subject to self-absorption.

1.1.3.2 Monochromators A monochromator serves to isolate a single atomic resonance line from the emission spectrum of the hollow cathode lamp or the electrodeless discharge lamp and to reject emissions from the atomizer. Ideally, the monochromator should be capable of isolating only the resonance line for the analyte and excluding all others. For some elements, such as copper, this is readily accomplished; for others, such as nickel, complete isolation is not achieved. The spectrum of copper is relatively uncluttered in the vicinity of the 324.8 nm resonance line. The nearest line in the copper spectrum is at 327.4 nm. The 232.0 nm nickel resonance line, on the other hand, is bounded with other nickel lines at 231.7 and 232.1 nm. Isolation of the 232.0 nm nickel resonance line is accomplished, in part, by narrowing the slits. The nonabsorbing line at 231.7 nm is particularly troublesome, and it is responsible for the nonlinearity of the calibration.

Most atomic absorption spectrometers employ gratings that are capable of spectral band passes at least as narrow as 0.1 nm. In addition, most atomic absorption spectrometers mechanically or electronically chop the output of the lamp. Such modulation allows electronic differentiation between the intensity of the resonance line and the emissions of the atomizer.

1.1.3.3 Detectors Most commercially produced atomic absorption spectrometers[3,4] employ a photomultiplier tube as the detector, although some of the more recently introduced instruments[5] make use of monolithic detectors incorporating photodiodes in combination with complementary metal oxide semiconductor (CMOS) charge amplifier arrays.

Resonance radiation photons from the hollow cathode lamp or from the electrodeless discharge lamp pass through the atomizer and the monochromator, and they enter the photomultiplier tube through an optical window. These photons interact with the photocathode material and bring about the ejection of photoelectrons. The photoelectrons are attracted to the first dynode stage under the influence of a potential gradient. The photoelectrons acquire sufficient kinetic energy to permit each to cause the ejection of several secondary electrons by interaction with the first dynode stage, where the process is repeated. Current amplifications of 10^6 or 10^7 are achieved in a ten-dynode-stage photomultiplier tube.

The photocathode material determines the photomultiplier tube's spectral responsiveness and sensitivity. The cesium–antimony cathodes demonstrate adequate sensitivity in the range of 200–700 nm. The trialkali photocathodes are of somewhat lower sensitivity, but the respond to photons of energies corresponding to the spectral range of 200–800 nm.

1.1.3.4 Atomization Devices The efficient and reproducible introduction of the analyte as an atomic vapor is, perhaps, the most difficult aspect of atomic absorption spectrometry. With the single possible exception of the cold vapor technique for mercury, atomization is achieved by thermal decomposition of solid, liquid, or gaseous form of the analyte. Both flame and electrothermal techniques have been used for this purpose.

Flames The production of an atomic vapor from a solution containing the analyte by flame atomization usually involves at least five steps: (1) nebulization, (2) droplet precipitation, (3) mixing of the aerosol mist with fuel and oxidant, (4) desolvation of the compounds in the aerosol, and (5) compound decomposition. The sample solution is drawn through a capillary into the nebulizer at a rate of between 3 and 6 mL/min by the venturi action of the flowing oxidant. Some of the liquid, ~5%, is dispersed as a fine aerosol mist. The larger droplets, ~90% of the aspirated material, precipitate and flow off to waste. The aerosol mist is mixed with the fuel and oxidant gases and emerges from the burner head into the flame. The aerosol particles undergo desolvation in the flame. The resulting solid particles are decomposed into their constituent atoms by the flame, thereby creating the environment of gaseous atoms necessary for atomic absorption.

Atomic absorption spectrometry is a comparative technique. Hence, it is necessary that atomization from the sample solutions and from the solutions of the reference standards take place to the same extent. Difference in viscosity between samples and standards will affect the aspiration rate and ultimately the number of gaseous analyte atoms. Differences in surface tension have a direct effect on nebulization efficiently, hence affect the number of gaseous analyte atoms. The anion of the analyte metal in the aerosolized compound affects the ease of decomposition into gaseous atoms. It is necessary, therefore, to match the reference standards as closely as possible to the matrices of the sample solutions, or to modify the sample matrices to match those of the standards.

Most commercial atomic absorption spectrometers accept both a 10 cm air–acetylene burner head and a 5 cm nitrous oxide–acetylene burner head. The air–acetylene burner head is standard for most atomic absorption measurements. It produces a temperature of 2300°C, and it shows good optical properties above 230 nm. The temperature of the nitrous oxide–acetylene flame is approximately 3000°C, which facilitates the determination of metals forming refractory oxides and eliminates some chemical and matrix interference's. Its optical properties are also good. Both the air–acetylene flame and the nitrous oxide–acetylene flame absorb appreciably below 200 nm. The air-entrained argon–hydrogen flame, which is more transparent at this wavelength, has been used to measure absorbances due to arsenic and selenium, the resonance lines for which are at 193.7 and 196.1 nm, respectively.

The selection of oxidant and fuel, the adjustment of the combustion ratio, and the location of the optical path in the flame are optimized on an analyte-by-analyte basis. These optimizations are included in Table 1.1.

Hydride Generators Antimony, arsenic, bismuth, germanium, selenium, tellurium, and tin form gaseous hydrides under easily controlled conditions. It is possible to atomize these elements from their hydrides, thereby improving sensitivity and reducing interferences in their respective determinations.

The hydride generator consists of a reaction vessel in which the hydride is formed, a means for adding the reducing agent to form the hydride, and a means for introducing the gaseous hydride into the atomic absorption spectrometer. Each manufacturer of atomic absorption spectrometers markets a hydride generator system that carries out these functions. In most systems, the atomization occurs in a heated quartz cell by thermal decomposition of the hydride. Thermal decomposition of the hydride has also been accomplished in the air-entrained argon–hydrogen flame.

Determinations made on the basis of hydride atomization are much more selective and sensitive than those made by aspirating solutions. The formation of the gaseous hydride and its subsequent introduction into the atomic absorption spectrometer leaves behind in the reaction vessel any matrix interferences the sample may have. The fraction of analyte actually atomized approaches 100% when hydride generation techniques are employed. In addition, absorbance by the flame is eliminated when atomization takes place in the heated quartz tube. All these factors enhance selectivity and sensitivity. The improved sensitivities of hydride generation techniques are summarized in Table 1.4.

Cold Vapor Generators Mercury is unique among the elements amenable to atomic absorption spectrometry; elemental mercury has an appreciable vapor pressure at room temperature (20 mg/m^3 at 25°C). Consequently, efficient introduction of atomic mercury vapor into a cell in the optical path of the atomic absorption spectrometer can be achieved with simple gas handling techniques.

TABLE 1.4 Comparison of Detection Limits for Flame Atomization with Hydride Generation Techniques

Element	Detection Limits (ppm)	
	Flame	Hydride
Antimony	45	0.15
Arsenic	150	0.03
Bismuth	30	0.03
Selenium	100	0.03
Tellurium	30	0.03
Tin	150	0.5

The cold vapor generator consists of a reaction vessel in which mercury is reduced to the elemental form, a means for adding the reducing agent, and a means for transporting the mercury vapor into the cell. Many hydride generating systems can also serve as cold vapor generators. In such applications, the cell is not heated. Mercury detection limits by cold vapor generation are some five orders of magnitude lower than those achieved by conventional flame atomization (i.e., 300 vs. 0.009 μg/L).

Several manufacturers[6–8] offer dedicated atomic absorption spectrophotometer–flow injection systems for automated mercury determinations. These systems which allow 25 mercury determinations per hour, boast detection limits below the parts-per-billion level. Zagatto et al.[9] have reviewed the applications of flow injection atomic spectrometry to environmental monitoring. Commercial instruments[10] are available both for fully automated flow injection–cold vapor and hydride atomic absorption spectrometry and for fully automated flow injection–conventional flame aspiration atomic absorption spectrometry. Direct flow injection–furnace coupling has also become available.[5]

Electrothermal Devices Electrically heated graphite tubes, which are more efficient than flames in atomizing the sample, are commonly used for producing atomic vapors. Only microliter quantities of the sample solution are required for the determination of analyte at concentrations two to three orders of magnitude below those measurable by conventional flame atomization. Typical working ranges are given in Table 1.1, and detection limits can be compared with the values contained in Table 1.3.

Electrothermal atomization takes place in an inert atmosphere to prevent oxidation of the carbon surface, and it usually involves four distinct, preprogrammed, time–temperature steps (for drying, ashing, atomizing, and "burning out"). The conditions of time and temperature depend both on the analyte and on the sample.

The drying step serves to remove solvent and to deposit the sample as a finely divided solid on the surface of the graphite tube. For aqueous solutions, this is usually accomplished at a temperature of 100°C. For 10 μL, 30 seconds usually suffices. For larger or smaller injections, 2.5 s/μL is usually adequate to dry the sample.

During the ashing step, volatilization of organic and low-boiling inorganic compounds occurs. The upper limit for this part of the atomization program is the highest temperature not resulting in the loss of analyte. The major function of the ashing step is to modify the solid residue on the graphite surface for reproducible and efficient atomization. Depending both on the composition of the dried film and on the specific properties of the analyte, the ashing step is carried out a temperatures between 100 and 1000°C for times ranging from 10 to 60 seconds.

The atomization step is carried out rapidly, in 5–10 seconds, at temperatures of 1000–3000°C. During this step, the thermal energy is sufficient to vaporize the ashed material and to form the atomic vapor of the analyte. It is during this step that the atomic absorbance is measured.

To remove any residual material before introduction of the next sample or standard, the electrothermal device is briefly heated to a temperature in excess of 2500°C. This "burning out" step is particularly important when atomic absorption signals from refractory analytes are being measured.

Electrothermal atomization has been accomplished using as the heating element various refractory metals (e.g., platinum, tantalum, tungsten), graphite or vitreous carbon, or metal-clad or -coated graphite fabricated into ribbons, rods, or tubes. The graphite furnace and the carbon rod atomizers are encountered more frequently than are those based on metal ribbon. In use, however, gradual oxidation leads to loss of reproducibility and sensitivity. It is necessary, therefore, to confirm calibration frequently, and to replace the furnace tube or rod when deterioration is observed. Depending on the sample matrix and atomization program, a conventional furnace tube has a typical useful life of 50–100 atomization cycles, and a pyrolytic coating would extend it by a factor of 3.

Programmable autosamplers, available for use in conjuction with electrothermal atomizers, make possible sequential determinations of up to eight elements in as many as six dozen samples with periodic recalibration from a maximum of nine standards.

1.1.4 Interferences in Atomic Absorption Spectrometry

Atomic absorption spectrometry is a comparative technique for the quantification of environmental contaminants. The technique is characterized by high sensitivity and selectivity. However, several properties of the resonance radiation, the matrix, the analyte, and the atomizer can contribute to interference with accurate comparisons of absorbances by the samples with those by the standards. These interferences can be classified as chemical, physical, and spectral, and their effects on the comparison can be minimized by matrix matching and/or background correction.

1.1.4.1 Chemical Interferences Chemical interferences occur when the analyte combines with components of the sample matrix to form compounds that undergo atomization differently from the standards. A typical example of chemical interference is that of phosphate from the matrix on the determination of calcium when calcium nitrate or calcium chloride standards are used. In the air–acetylene flame, calcium phosphate is converted to calcium pyrophosphate, which is more thermally stable than is calcium nitrate or calcium chloride. Hence, a smaller fraction of free calcium atoms are formed from the samples than are formed from the standards. The absorbance from the former is correspondingly lower. Consequently, the result will show a negative bias.

The effects of chemical interference can be minimized by adding chemicals that yield easily atomized analytes to both the samples and standards. In the case of the phosphate interference on the determination of calcium in the air–acetylene flame, the addition of lanthanum chloride will form lanthanum phosphate, thereby releasing calcium for atomization. Addition of EDTA to both samples and standards similarly eliminated chemical interferences in the determination of lead.[11]

Some success has been achieved by using higher atomization temperatures for minimizing the effects of chemical interferences (e.g., 2900°C for the nitrous oxide–acetylene flame vs. 2200°C for the air–acetylene flame).

The suppression of analyte ionization by matrix elements is another potential chemical interference. At low concentrations, a significant number of, for example, potassium atoms would undergo ionization during flame or furnace atomization. The population of ground state atoms available for the absorption of the resonance radiation would be reduced correspondingly. In the presence of large amounts of sodium, the thermal ionization of potassium would be suppressed. Hence, it is possible to obtain high results when aqueous potassium chloride standards are used to measure serum potassium levels. To equalize the extent of ionization, an ionization buffer of sodium chloride could be added to both samples and standards. The use of ionization buffers is recommended for the determination of alkali and alkaline earth metals, especially when high temperature atomization is used to overcome the phosphate interference on the latter.

The loss of volatile analyte compounds during the drying or ashing step in electrothermal atomization also constitutes chemical interference. In blood lead determinations, low results could be obtained from the loss of lead chloride formed by interaction with the matrix prior to the atomization step. Interferences of this type should be addressed by means of standard addition and/or matrix modification. The former is described in Section 1.1.5.2, and the latter is accomplished by chemical means to convert the analyte to a less volatile compound. The chemical modifier is added both to standards and samples.

1.1.4.2 Physical Interferences Physical interferences are the result of differences between the physical properties of the standards and the samples that are of sufficient magnitude to affect the nebulization/atomization process. Both flame and furnace atomization suffer from interferences of this kind. Matrix matching, matrix modification, and background correction are often employed to minimize the effects of physical interferences. The method of standard additions described in Section 1.1.5.2 is valuable in minimizing the effects of these interferences.

In conventional flame atomic absorption spectrometry, the viscosity and surface tension of the sample solutions are influenced by the matrix acid, dissolved solids, and, in the case of body fluids, the protein content. Viscosity and surface tension influence, in turn, the aspiration, nebulization, and atomization efficiencies of the analyte. Because atomic absorption spectrometry is a comparative technique, differences between the sample and standard matrices could lead to errors when absorbances are compared. Matrix matching or matrix modification by solvent extraction, sample dilution, and/or addition of specific chemical reagents to both samples and standards is a convenient technique to prevent the introduction of many of these errors.

In both flame and furnace atomization, the formation of a particulate cloud or combustion product vapor could cause absorption or scatter of the resonance radiation by materials other than the analyte. This would give rise to false absorbance signals. Compensation for false absorption signals is made by background correction rather than by matrix matching or matrix modification. So important is proper background correction, that Veillon[12] has cast doubt on the validity of all urinary chromium concentrations made by electrothermal atomic absorption spectrometry using the deuterium lamp for this purpose.

Background Correction Background correction is made by measuring the attenuation of both the resonance line and the reference radiation as they transverse the atomizer. The reference radiation may be a continuum from an auxiliary lamp, or Zeeman-shifted radiation from the atomic line. Attenuation of the resonance radiation intensity is attributed to both the analyte and the background, while the attenuation of the reference radiation is attributed only to the background. The correction for background absorption of the resonance line is obtained from the attenuation of the reference radiation at a wavelength very near the resonance line. Background correction with the deuterium lamp is applicable to measurements made at wavelengths below 300 nm; Zeeman effect background corrections have a wider range of applicability, and Zeeman background correction has become the standard for most atomic absorption spectrophotometers.

1.1.4.3 Spectral Interferences Spectral interferences occur when radiation other than the resonance line of the analyte contributes to or detracts from the absorbance. Modulation of the incident resonance line from the hollow cathode lamp prevents emissions from atomizer, matrix, or analyte from affecting the absorbance. Because of the narrowness of the resonance line ($\sim 10^{-5}$ nm), there is little likelihood of absorbance by substances other than the analyte atoms, provided of course that adequate background correction for molecular absorption is achieved. On the chance atomic that absorption of an interferant does occur within the narrow spectral line of the incident radiation, alternate wavelengths can be selected for the measurements.

1.1.4.4 Identification of Interferences The split and spike approach is useful for identifying interferences particularly those due to the matrix. The following steps can be used to incorporate this approach into the laboratory's quality assurance/quality control program:

1. Dilute an aliquot of the sample with water, 1:4.
2. Spke a second aliquot with a known mass of analyte and dilute it to the same volume as the first.
3. Measure the aborbances of the two diluted aliquots and the original sample.

In the absence of interferences, the concentration of analyte found in the original should agree within ±10% of that found in the first diluted aliquot, and the recovery of the spike added to the second aliquot should be 90–110% relative to the first aliquot. Spike recovery is determined as:

$$R = \frac{100\ C_m(V_s + V_a) - (C_s V_s)}{\text{mass of spike}}$$

where C and V refer to the concentrations and volumes of the mixtures, m, the sample, s, and the addition, a. If any significant interferences are identified, the sample should be evaluated by the method of standard additions as described in Section 1.1.5.2.

1.1.5 Calibration

The objective in making atomic absorption measurements is to relate the absorbance of a sample to its analyte content. This is accomplished by comparing absorbances (sample vs. standards) and assuming that the same absorbance–concentration relationship holds for both standards and samples. Two approaches are available for establishing the analyte content of the samples: direct comparison and standard additions.

1.1.5.1 Direct Comparison The relationship developed in Section 1.1.2 shows a direct proportionality between absorbance and the number of gaseous atoms in the system. If the atomization efficiency is constant, this proportionality to the solutions containing the analyte allows its quantification. Ideally, this linear relationship should be demonstrated when absorbances A_1, A_2, A_3, . . . are plotted against concentrations of analyte in the standard solutions C_1, C_2, C_3, ... If proper attention has been given to minimizing interferences, such a calibration curve can be used to determine the concentrations of analyte in the samples C_x, C_y, C_z, . . . from their absorbances A_x, A_y, A_z, ...

Absorbance–concentration plots are rarely linear over a wide range of concentrations. Typically, such plots show curvature toward the concentration axis at the higher concentrations. The extent of curvature and the concentration at which it begins depend on the amount of parasitic radiation stimulating the detector. Even with a narrow band pass, nearby lines from the hollow cathode lamp transverse the atomizer. Since each of these lines is absorbed differently, or not at all in the case of nonabsorbing lines, absorbance at increasing concentrations deviates from linearity. Although current instrumentation is capable of curvature correction, it is equally feasible to dilute the sample solutions so that their absorbances fall within the linear portion of the calibration curve.

1.1.5.2 Standard Additions Although it appears to be less convenient than direct comparison, standard additions is the method of choice when only a few samples are to be evaluated, when the samples differ from each other in matrix, and/or when the samples suffer from unidentified matrix interferences. The method of standard additions is carried out by (1) dividing the sample into several—at least four—aliquots, (2) adding to all but the first aliquot increasing amounts of analyte, (3) diluting all to the same final volume, (4) measuring the absorbances, and (5) plotting the absorbance against the amount of analyte added, as shown in Figure 1.1. The amount of analyte present in the sample is obtained by extrapolation beyond the zero addition. For extrapolation to succeed, the addition may not exceed the linear portion of the curve.

Alternatively, the concentration of analyte in the sample, C_s, can be computed from:

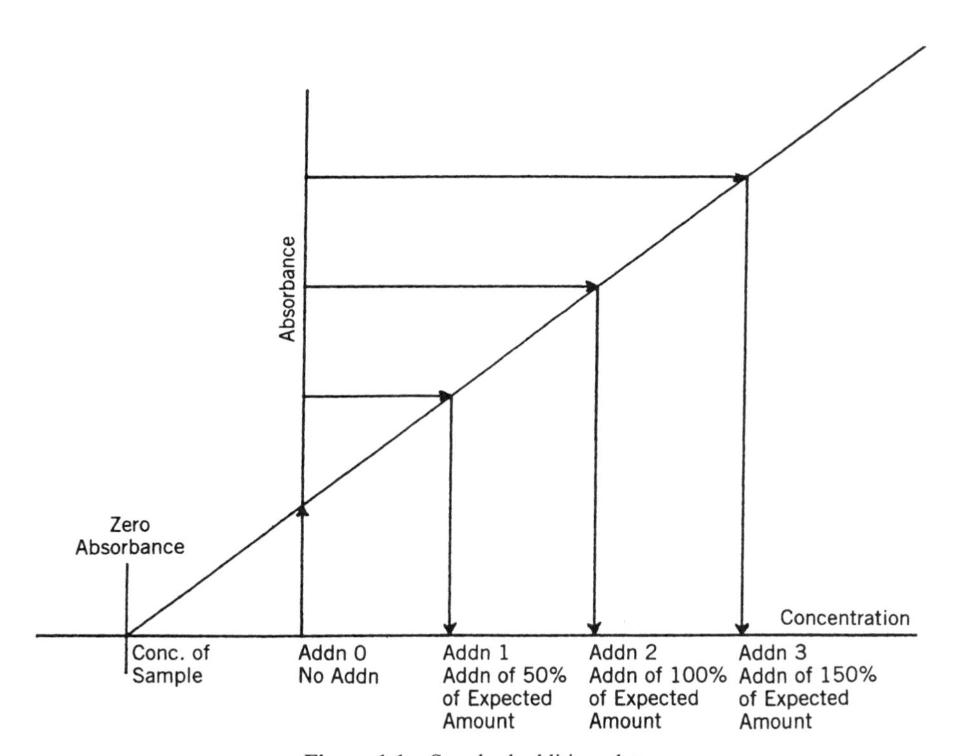

Figure 1.1 Standard addition plot.

$$C_s = \frac{A_s V_a C_a}{(A_m V_a + A_m V_s + A_s V_s}$$

where C, A, and V refer to concentrations, absorbances, and volumes of the original sample, s, the addition, a, and the mixture, m.

The computations are carried out by the microprocessor controlling the atomic absorption spectrometer. The method of standard additions is less accurate than direct comparison; it is necessary, however, when matrix interferences are encountered.

1.2 INDUCTIVELY COUPLED PLASMA–ATOMIC EMISSION SPECTROMETRY

Atomic emission spectrometry preceded atomic absorption spectrometry but fell into disfavor when the newer technique was shown to be superior in speed, simplicity, selectivity, sensitivity, and accuracy. It was not long, however, before the electronic revolution coupled with the development of plasma excitation sources and computerized signal processing rejuvenated atomic emission spectrometry. Atomic emission spectrometry and atomic absorption spectrometry have become complementary approaches to regulatory compliance monitoring.

1.2.1 Scope and Sensitivity of Inductively Coupled Plasma–Atomic Emission Spectrometry

Inspection of Table 1.3 reveals that inductively coupled plasma–atomic emission spectrometry is similar to flame atomic absorption spectrometry in scope and sensitivity. Although some atomic absorption spectrophotometers[5,13] with automated electrothermal atomization systems are capable of measuring up to five elements simultaneously, inductively coupled plasma–atomic emission spectrometry is more amenable to simultaneous multielement determinations. In addition, the higher temperature of the plasma is an important factor in reducing the chemical interferences associated with atomic absorption spectrometry. However, the sensitivity of electrothermal atomization atomic absorption spectrometry is usually superior to that of inductively coupled plasma–atomic emission spectrometry.

1.2.2 Theory of Inductively Coupled Plasma–Atomic Emission Spectrometry

Inductively coupled plasma–atomic emission spectrometry is based on excitation of analyte atoms in the argon plasma and measurement of the characteristic emissions from excited analyte atoms or ions as a consequence of decay to lower energy states. The emissions correspond to specific quantum changes in the electronic energies of the excited analyte atoms or ions. The wavelengths of the emitted radiations are characteristic line spectra.

The intensity I of an atomic spectral line is:

$$I = N_2 h\nu g_2 b \frac{A}{B} e^{-\Delta E/kT}$$

where N_2 is the number of analyte atoms or ions in excited state 2, h is Planck's constant ($= 6.6 \times 10^{-34}$ J/s), ν is the frequency of the emitted radiation, d is the diameter of the discharge in the direction of the detector, and A and B are constants for the transition probability and temperature conditions, respectively. From the Boltzmann distribution introduced in Section 1.1.2 [i.e., $N_1/N_2 = (g_1/g_2)e^{-\Delta E/kT}$], it follows that the intensity of a line in the emission spectrum of the analyte is proportional to the number of analyte atoms:

$$I \propto N_1$$

To make use of this proportionality for quantifying the analyte, it is necessary to compare emission intensities of samples and standards under the same excitation conditions at a wavelength free of spectral interferences. Inductively coupled plasma–atomic emission spectrometry, like atomic absorption spectrometry, is a comparative rather than an absolute measurement technique.

1.2.3 Requirements for Inductively Coupled Plasma–Atomic Emission Spectrometry

The measurement of atomic emission requires a plasma for exciting the analyte atoms, a means for introducing the sample or standard into the plasma, a monochromator for resolving the atomic or ionic emissions into discrete lines, and a detector for measuring the intensity of the spectral line.

1.2.3.1 Excitation Sources The excitation source used for inductively coupled plasma–atomic emission spectrometry is usually a torch constructed of three concentric quartz tubes positioned axially within an induction coil. Argon flowing through the outermost tube is initially made conductive by electrical discharge. Under the influence of the oscillating electromagnetic field of the induction coil, electrons from the primary ion pairs are accelerated until they induce secondary ionizations. Stabilization of the ionic population produces a plasma with a temperature of 10,000 K. Argon flowing through the middle quartz tube controls the position of the plasma within the torch, and the argon flowing through the central tube of the torch serves to transport the samples and standards into the plasma.

The primary (coolant, or plasma) flow of argon through the plasma torch must be sufficiently high (~ 15 L/min) to maintain the walls if the quartz tubes at temperatures below the melting point. In normal operation, the auxiliary argon flow for positioning the plasma is approximately 1 L/min. The sample or nebulizer argon flow is also approximately 1 L/min. Argon consumption is high, and the control of argon flow rates is essential for reproducible plasmas.

The radio frequency (rf) generators driving the induction coils operate at power levels and frequencies of approximately 1 kW and 40 MHz, respectively. The induction coils that transfer the radio frequency power to the plasma are often fabricated from copper, and they are water cooled.

An inductively coupled plasma torch is shown schematically in Figure 1.2.

1.2.3.2 Introduction of Sample/Standard The introduction of samples or standards into the inductively coupled plasma corresponds to aspiration into the atomic absorption flame. While some designs are able to aspirate the solution containing the sample or standard, a peristaltic pump is often employed for transporting the solution to the nebulizer. The use of the peristaltic pump also produces a constant flow rate that is independent of solution parameters such as surface tension and viscosity.

Contact between the high speed nebulizer argon flow and the solution of sample or standard causes the liquid to form an aerosol. This has been achieved with both concentric and cross-flow nebulizer designs as well as with the Babington design and its subsequent V-groove modification. Considerations of solids contents and pH are important for avoiding clogging and/or corroding the nebulizer. Ultrasonic nebulization, electrothermal atomization, and hydride generation have also been successfully employed for the introduction of samples and standards to the plasma torch.

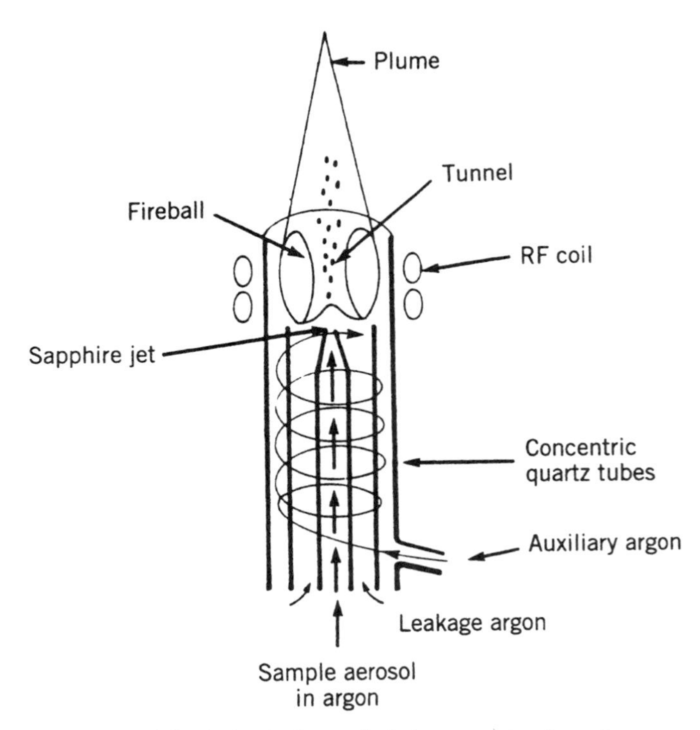

Figure 1.2 Inductively coupled plasma torch schematic.

Only the smallest droplets of the aerosol from the nebulizer are suitable for injection into the plasma, and their particle size distribution is important for reducing baseline noise in radially viewed plasmas. Large droplets (diameters $\geq$ 10 μm) are condensed from the aerosol in the spray chamber. Less than 5% of the sample/standard introduced to the nebulizer is injected into the plasma; more than 95% is drained away to waste from the spray chamber. In the spray chamber, pulses in the flow rate from the peristaltic pump are smoothed, and the sample/standard is injected into the plasma at a constant rate.

1.2.3.3 Monochromators/Polychromators Excitations of and emissions from the analytes occur in the inductively coupled plasma. Arguments on whether these emissions are best viewed radially or axially tend to favor the latter mode. Identifications and quantifications of analytes requires resolution of the emitted radiations into their characteristic lines and measurement of their intensities at the corresponding wavelengths with a high resolution optical system.

Diffraction gratings are used in a variety of echelle[14] and Czerny–Turner[15–17] optical systems to disperse the emitted radiations. The grating functions as a monochromator for sequential measurements and as a polychromator for simultaneous measurements on the emitted radiations. The latter are more demanding of the optical system. The emitted radiations are resolved, and the emission lines are focused on the exit slit(s). For sequential measurements of emission line intensities, rotation of the grating or other component of the optical system allows the region of spectral interest to pass through the exit slit and stimulate the detector. Scanning or slewing from one emission line to another makes possible sequential measurements of their intensities. For simultaneous measurements, several predetermined regions of spectral interest are individually focused on separate exit slits, behind each of which is located a detector. With exact optical alignments, simultaneous measurements on the intensities of several emission lines are possible. Table 1.5 lists some of the wavelengths, detection limits,[18] and standard working ranges[19] for inductively coupled plasma–atomic emission spectrometry. In comparison to the wavelengths and working ranges for atomic absorption spectrometry listed in Table 1.1, those for inductively coupled–atomic emission spectrometry reflect greater demands for spectral purity and wider dynamic ranges.

1.2.3.4 Detectors Although detection based on charge injection or charge collection devices has been applied successfully in inductively coupled plasma–atomic emission spectrometry,[18,20,21] most commercially produced instruments now employ the photomultiplier tube as the detector (see Section 1.1.3.3).

1.2.4 Interferences in Inductively Coupled Plasma–Atomic Emission Spectrometry

The high temperature of the inductively coupled plasma essentially eliminate chemical interferences, and the peristaltic pump and spray chamber included in the sam-

TABLE 1.5 Detection Limits[18] and Standard Working Ranges[19] for Inductively Coupled Plasma–Atomic Emission Spectrometry

Element	Wavelength (nm)	Detection Limit (μg/L)	Working Range (mg/L)
Ag	328.068	0.4	0.003–150
Al	308.215	4	0.025–500
As	189.042	2	0.050–250
B	249.678	0.5	0.006–150
Ba	493.409		0.001–100
Be	313.042	0.08	0.001–50
Bi	223.061	2	0.030–500
Ca	393.366		0.001–50
Cd	226.502	0.2	0.002–150
Co	228.616	0.5	0.002–150
Cr	267.716	0.4	0.005–150
Cu	324.754	0.5	0.002–150
Fe	259.940	0.3	0.005–150
K	766.491	2	0.200–1000
Mg	279.553	0.05	0.001–50
Mn	257.610	0.1	0.001–100
Mo	202.030	0.7	0.005–200
Na	588.995		0.010–100
Ni	231.604	0.7	0.010–200
Pb	220.353	0.8	0.025–200
Sb	206.838	2	0.050–200
Se	196.026	2	0.050–250
Sn	189.989		0.020–250
Ti	337.000	0.2	
Tl	190.864	2	0.050–500
V	292.402	0.4	0.002–200
Zn	213.856	0.1	0.004–100

ple introduction system do much to minimize the physical interferences described in Section 1.1.4. To assure the reproducibility of sample introduction, nebulizer clogging must be avoided. Since, however, emission spectra from complex samples contain many lines, possibilities for spectral interferences require serious considerations.

1.2.4.1 Spectral Interferences Spectral interferences are caused by overlap on the line of interest by a spectral line from another element and/or stray light contributions to the background from the emission lines of other elements present at high concentrations in the sample. The effects of these interferences with the spectral line are minimized by the high resolution optical system of the spectrometer, the software-controlled selection of an alternate emission line, and the software-controlled application of interelement correction factors. Background corrections are made by "peak/off peak" measurements of the emission intensities. Although considered tedious by some, the method of standard additions described in Section 1.1.5.2 is an appropriate approach to the identification of and compensation for these interferences. The method of standard additions is also appropriate for correcting matrix interferences.

Some interferences from 100 ppm titanium, vanadium, chromium, manganese, iron, nickel, copper on the emission lines of other elements and the "peak /off peak" wavelengths for background corrections are listed in Table 1.6.[22,23]

1.2.4.2 Identification of Interferences The split and spike approach described in Section 1.1.4.4 is applicable to the identification of interferences in determinations made by inductively coupled plasma–atomic emission spectrometry. Interferences as well as instrument drift and other deviations from normal operating conditions are also identified by failures to achieve results within the control limits for the check standards described in Section 4.6.3.3.

1.2.5 Calibration

Like atomic absorption spectrometry, inductively coupled plasma–atomic emission spectrometry is a comparative rather than an absolute measurement technique. However, unlike atomic absorption spectrometry, calibration of the inductively coupled plasma–atomic emission spectrometer is two-dimensional; thus it is necessary to determine relationships between wavelengths for identification and detector response for quantification. In addition, wavelengths for background corrections and factors for interelement corrections must be established. Also, the flows of argon to the torch and the flow of sample to the nebulizer must be standardized. Inductively

TABLE 1.6 Some Apparent Concentrations of Analyte from 100 mg/L Interferant, and "Peak/off Peak" Wavelengths for Background Corrections

	Wavelengths (nm)[23]		Interferant[22]						
Analyte	Peak	Off Peak	Ti	V	Cr	Mn	Fe	Ni	Cu
As	193.695	193.717		1.10	0.44				
Ba	455.403	455.342							
Be	234.861	234.830	0.04	0.05					
Cd	226.502	226.480				0.03	0.02		
Cr	267.716	267.683		0.04		0.04	0.01		
Ni	231.604	231.577							
Pb	220.353	220.326							
Sb	217.581	217.560	0.25	0.45	2.90		0.08		
Se	196.026	196.042							
		196.007					0.09		
Tl	190.801	190.784							
V	292.401	292.373	0.02		0.05		0.01		
Zn	213.856	213.883						0.29	0.14

coupled plasma–atomic emission spectrometers are calibrated with multielement standards while the instrument is under control of the computer. First the plasma is optimized and tuned. The instrument is then calibrated with multielement standard solutions. Finally the calibration is confirmed with laboratory performance check solutions. Calibration should be reconfirmed with the laboratory performance check solution after every tenth sample. When the results for the laboratory performance check solution are not within 95–105% of the reference values, recalibration of the instrument is required.

1.3 INDUCTIVELY COUPLED PLASMA–MASS SPECTROMETRY

Inductively coupled plasma–mass spectrometry (ICP-MS) is not inductively coupled plasma–atomic emission spectrometry with a new detector, nor is it mass spectrometry with a new ion source. Inductively coupled plasma–mass spectrometry is a sensitive and selective technique for rapid, multielement determinations made possible by successfully interfacing the plasma torch and sample introduction system of inductively coupled plasma atomic emission spectrometry with the ion optics, the quadrupole mass analyzer, and the channel electron multiplier detector of mass spectrometry.

1.3.1 Scope and Sensitivity of Inductively Coupled Plasma–Mass Spectrometry

Some six dozen elements are amenable to identification and quantification by inductively coupled plasma–mass spectrometry, a technique that is superior in sensitivity to electrothermal atomic absorption spectrometry, as indicated earlier (Table 1.3). Some of the elements routinely identified and quantified with inductively coupled plasma–mass spectrometry are listed in Table 1.7.

1.3.2 Theory of Inductively Coupled Plasma–Mass Spectrometry

One of the factors that determines the detector response to an element in the sample or standard solution is the population of its singly charged species in the plasma. This population varies with the ionization potential of the element[24]; that is,

$$\frac{N_{ij}N_e}{N_{aj}} = \left[\frac{(2\pi m_e kT)^{2/3} 2z_{ij}}{h^3 z_{aj}}\right] e^{-ej/kT}$$

where N_{ij} = concentration of ionic species j
N_e = concentration of free electrons
N_{aj} = concentration of atomic species j
m_e = mass of the electron
h = Planck's constant
k = Boltzmann's constant
T = ionization temperature
z_{ij} = partition function of ionic species j
z_{aj} = partition function of atomic species j
ej = ionization potential of atomic species j

TABLE 1.7 Elements Routinely Identified and Quantified with Inductively coupled Plasma–Mass Spectrometry

Analyte	Recommended Ion	Detection Limit (μg/L)[a]
Aluminum	$^{27}Al^+$	0.05
Antimony	$^{123}Sb^+$	0.08
Arsenic	$^{75}As^+$	0.9
Barium	$^{137}Ba^+$	0.5
Beryllium	$^{9}Be^+$	0.1
Cadmium	$^{111}Cd^+$	0.1
Chromium	$^{52}Cr^+$	0.07
Cobalt	$^{59}Co^+$	0.03
Copper	$^{63}Cu^+$	0.03
Lead	$^{206}Pb^+$	0.08
Lead	$^{207}Pb^+$	0.08
Lead	$^{208}Pb^+$	0.08
Manganese	$^{55}Mn^+$	0.1
Mercury	$^{202}Hg^+$	1
Molybdenum	$^{98}Mo^+$	0.1
Nickel	$^{60}Ni^+$	0.2
Selenium	$^{82}Se^+$	5
Silver	$^{107}Ag^+$	0.05
Thallium	$^{205}Tl^+$	0.09
Thorium	$^{232}Th^+$	0.03
Uranium	$^{238}U^+$	0.02
Vanadium	$^{51}V^+$	0.02
Zinc	$^{66}Zn^+$	0.2

[a]These scanning mode detection limits can be lowered by a factor of 5 or more in the selective ion mode with dwell times of ~1 s/amu.

Olesik[25] described the factors involved in converting samples to signals and explained how they influence results obtained by inductively coupled plasma–mass spectrometry. In addition to plasma dynamics, aerosol generation and transport, and ion production and transport, he considered effects of solvent, background and spectral overlap, and dissolved solids. He proposed that "A minuscule fraction of the analyte ions (often less than one in a million) reach the detector of the mass spectrometer." In the ideal situation, the relationship between analyte concentration in the sample and detector response is linear. This is not the case in the real situation. Deviations from linearity occur, and interferences are encountered. These are described in Section 1.3.4.

1.3.3 Requirements for Inductively Coupled Plasma–Mass Spectrometry

In inductively coupled plasma–mass spectrometry, standard or sample solution introduced into the argon plasma by pneumatic nebulization undergoes ionization.

The ions are extracted from the plasma through a differentially pumped vacuum interface into the quadrupole mass analyzer, where they are separated on the basis of mass-to-charge ratios. Resolution of at least 1 amu through mass 250 is required. The ions transmitted through the quadrupole are detected by a charge multiplier detector. A recent review has described some of the instrumentation currently available for inductively coupled plasma–mass spectrometry.[26]

1.3.3.1 ***Sample Introduction/Sample Ionization*** The peristaltic pumps and nebulizers for inductively coupled plasma–mass spectrometry are similar to those used for inductively coupled plasma–atomic emission spectrometry. The spray chambers for the former are frequently cooled to condense water vapor from the aerosol entering the plasma. Decomposition of the water in the plasma and subsequent recombination of the oxygen with the argon or components of the sample matrix may yield products that have overlaps with some of the analytes (e.g., both $^{40}Ar^{16}O^+$ and $^{56}Fe^+$ have mass-to-charge ratios of 56). While software could compensate for this overlap with correction factors derived from other iron isotopes, their low abundances would make such correction difficult at low concentrations of iron (^{54}Fe ~ 5.8%, ^{57}Fe ~ 2.2%, ^{58}Fe < 0.3%, etc.).

1.3.3.2 ***Ion Source–Mass Analyzer Interface*** The plasma torch is usually positioned perpendicular to the quadrupole mass analyzer interface. A diagram of the interface between the plasma ion source and the quadrupole mass analyzer is shown in Figure 1.3. A small fraction of the ions from the plasma pass through the water-cooled sampler cone, and an even smaller fraction pass through the skimmer cone into the high vacuum of the ion optics region of the mass spectrometer. Here they are focused as an ion beam into the quadrupole. The ion optics are optimized for a range of masses matching those of the analyte ions.

1.3.3.3 ***Mass Analyzer*** In inductively coupled plasma–mass spectrometry, the quadrupole mass analyzer performs a function analogous to that of the monochromator in inductively coupled plasma–atomic emission spectrometry. Only ions having a specific mass-to-charge ratio pass through the analyzer. Ions with other mass-to-charge ratios are removed by collision with one of the poles. Mass scanning is achieved by varying both the dc potentials and the rf ac potentials applied to the poles.

1.3.3.4 ***Detector*** Ions having a specific mass-to-charge ratio exit the mass analyzer and interact with the channel electron multiplier (CEM) detector. The CEM performs a function analogous to that of the photomultiplier tube detector used in inductively coupled plasma–atomic emission spectrometry. The beam of ions having a specific mass-to-charge ratio (analogous to a beam of monochromatic light) impacts on the first dynode and brings about the ejection of electrons, which are attracted to the next dynode stage under the influence of a potential gradient. These primary electrons acquire sufficient kinetic energy to initiate the ejection of sec-

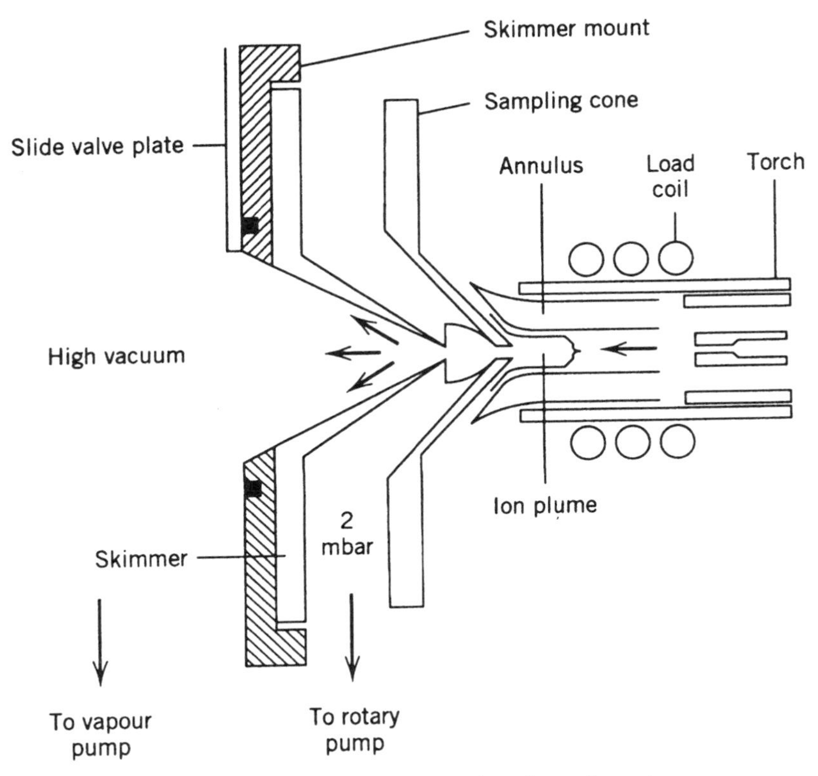

Figure 1.3 ICP–quadrupole MS interface schematic.

ondary electrons. The process continues until the electrons are captured by the collector, where an electrical signal proportional to the number of analyte atoms in the original solution is generated. Maintaining the proportionality between analyte atoms in the solution and electrons collected in the CEM is necessary for quantification.

1.3.3.5 Computer The computer serves many functions in the inductively coupled plasma–mass spectrometer: operation and calibration of the instrument, correction for background, corrections for both interelement and isobaric interferences, and quantification of the analytes. Menu-driven software provides for the transport of data to spreadsheets and databases.

1.3.4 Interferences in Inductively Coupled Plasma–Mass Spectrometry

The interferences in inductively coupled plasma–mass spectrometry parallel those in inductively coupled plasma–atomic emission spectrometry. They can be classified as either spectral or physical.

1.3.4.1 Spectral Interferences Isobaric polyatomic ion interferences occur when polyatomic ions formed in the plasma or at the ion source–mass analyzer interface have the same nominal mass-to-charge ratio, which cannot be resolved by the spectrometer, as that of the analyte. In addition to the $^{40}Ar^{16}O^+$ interference on $^{56}Fe^+$ cited Section 1.3.3.1, other common interferences have been identified. These are listed in Table 1.8.

In addition to interferences from isobaric polyatomic ions, one sometimes encounters isobaric elemental interferences from elemental ions having the same nominal mass-to-charge ratio, which cannot be resolved by the spectrometer, as that of the analyte. Examples include $^{54}Fe^+$ and $^{54}Cr^+$. Almost all analytes have at least one naturally occurring isotope free of isobaric elemental interferences.

Of more concern are interferences due to abundance sensitivity or contributions to the intensity of the analyte peak signal from high concentrations of isotopes having adjacent mass numbers (e.g., low concentrations of $^{51}V^+$ analyte and high con-

TABLE 1.8 Some Common Isobaric Polyatomic Ionic Interferences in Inductively Coupled Plasma–Mass Spectrometry

Interferant	Mass	Analyte	Interferant	Mass	Analyte
$^{14}N^1H^+$	15		$^{81}BrH^+$	82	$^{82}Se^+$
$^{16}O^1H^+$	17		$^{79}BrO^+$	95	$^{95}Mo^+$
$^{16}O^1H_2{}^+$	18		$^{81}BrO^+$	97	$^{97}Mo^+$
$^{12}C_2{}^+$	24		$^{81}BrOH^+$	98	$^{98}Mo^+$
$^{12}C^{14}N^+$	26		$Ar^{81}Br^+$	121	$^{121}Sb^+$
$^{12}C^{16}O^+$	28		$^{35}ClO^+$	51	$^{51}V^+$
$^{14}N_2{}^+$	28		$^{35}Cl^{16}O^1H^+$	52	$^{52}Cr^+$
$^{14}N_{21}H^+$	29		$^{37}Cl^{16}O^+$	53	$^{53}Cr^+$
$^{14}N^{16}O^+$	30		$^{37}Cl^{16}O^1H^+$	54	$^{54}Cr^+$
$^{14}N^{16}O^1H^+$	31		$^{40}Ar^{35}Cl^+$	75	$^{75}As^+$
$^{16}O_2{}^+$	32		$^{37}Ar^{35}Cl^+$	77	$^{77}Se^+$
$^{16}O_{21}H^+$	33		$^{32}S^{16}O^+$	48	
$^{36}Ar^1H^+$	37		$^{32}S^{16}O^1H^+$	49	
$^{38}Ar^1H^+$	39		$^{34}S^{16}O^+$	50	$^{50}V^+$
$^{40}Ar^1H^+$	41		$^{34}S^{16}O^+$	50	$^{50}Cr^+$
$^{12}C^{16}O_2{}^+$	44		$^{34}S^{16}O^1H^+$	51	$^{51}V^+$
$^{12}C^{16}O_2{}^1H^+$	45	$^{45}Sc^+$	$^{32}S^{16}O_2{}^+$	64	$^{64}Zn^+$
$^{40}Ar^{12}C^+$	52	$^{52}Cr^+$	$^{32}S_2{}^+$	64	$^{64}Zn^+$
$^{36}Ar^{16}O^+$	52	$^{52}Cr^+$	$^{40}Ar^{32}S^+$	72	
$^{40}Ar^{14}N^+$	54	$^{54}Cr^+$	$^{42}Ar^{32}S^+$	74	
$^{40}Ar^{14}N^1H^+$	55	$^{55}Mn^+$	$^{31}P^{16}O^+$	47	
$^{40}Ar^{16}O^+$	56		$^{31}P^{16}O^1H^+$	48	
$^{40}Ar^{16}O^1H^+$	57		$^{31}P^{16}O_2{}^+$	63	$^{63}Cu^+$
$^{40}Ar^{36}Ar^+$	76	$^{76}Se^+$	$^{40}Ar^{31}P^+$	71	
$^{40}Ar^{38}Ar^+$	78	$^{78}Se^+$	$^{40}Ar^{23}Na^+$	63	$^{63}Cu^+$
$^{40}Ar_2{}^+$	80	$^{80}Se^+$	$^{40}Ar^{39}K^+$	79	

centrations of $^{52}Cr^{+}$ interferant). The concentration of $^{51}V^{+}$ is computer corrected for background and for these interelement effects with empirical elemental equations derived from the calibration data and the mass spectrum of the sample. A typical example for vanadium[27] is:

$$C = 1.000\,{}^{51}C - 3.127\,{}^{53}C - 0.113\,{}^{52}C$$

where ^{51}C = uncorrected concentration mass 51
^{53}C = corrected concentration of mass 53
^{52}C = concentration of mass 52 corrected for isobaric interferences from $^{35}Cl^{16}O^{1}H^{+}$

Similar elemental equations are established for each analyte.

1.3.4.2 Physical Interferences Physical interferences in inductively coupled plasma–mass spectrometry focus on the physical processes associated with the generation and transport of the aerosol and the production and transport of the ions. Factors such as viscosity, surface tension, and dissolved solids which result in matrix differences between the samples and the standards, are responsible for physical interferences. Internal standardization may be used to compensate for physical interferences in a manner analogous to standard addition with the added requirement that the analytical behavior of the internal standards match that of the analytes.

1.3.5 Calibration

Calibration of the spectrometer is two-dimensional (i.e., by mass for identification of analytes and by intensity or number of pulses for quantification of analytes). Calibration by mass is accomplished with a tuning solution containing beryllium, cobalt, indium, lead, and magnesium each at 100 μg/L in 1% (v/v) nitric acid. The mass resolution of the spectrometer is adjusted to produce peak widths of less than 1 amu at 5% peak height using the peaks for ^{24}Mg, ^{25}Mg, and ^{26}Mg at the low mass end of the spectrum and those for ^{206}Pb, ^{207}Pb, and ^{208}Pb at the high mass end. Stability of the mass calibration is confirmed when pentuplicate mass spectra of the tuning solution show relative standard deviations of less than 5% for the absolute signals from all analytes. To calibrate by intensity, mass spectra of multielement standards in 1% (v/v) nitric acid and a calibration blank, 1% (v/v) nitric acid, are recorded in triplicate, and the means of the integrated signals are used to establish relationships between concentration and pulses counted.

Internal standardization, to allow corrections for instruments drift and matrix interferences, is accomplished by adding to all samples, standards, and blanks an appropriate internal standard (bismuth, indium, scandium, terbium or yttrium).

1.4 REGULATORY AGENCIES' REQUIREMENTS AND RECOMMENDATIONS FOR COMPLIANCE MONITORING BY ATOMIC SPECTROMETRY

To protect public health and environmental quality, many federal, state, and municipal agencies require, or at least recommend, measurements of trace element concentrations in a wide variety of matrices. In many cases, atomic spectrometry is required or recommended for these measurements. The elements and matrices range from chromium in the urine of electroplaters to sodium in peanut butter destined for school lunch programs.

Among the acts promulgated to protect public health and environmental quality are the Comprehensive Environmental Response, Compensation and Liability Act (CERCLA: PL 96-510), the Superfund Authorization and Reauthorization Act (SARA: PL 99-499), the Resource Conservation and Recovery Act (RCRA: PL 94-580), the Safe Drinking Water Act (SDWA: PL 93-523), the Water Pollution Control Act (WPCA: PL 92-500), the Clean Water Act (CWA: PL 95-217), and the Water Quality Act (WQA: PL 100-4).

1.4.1 Comprehensive Environmental Response, Compensation and Liability Act (CERCLA) and Superfund Authorization and Reauthorization Act (SARA)

The Comprehensive Environmental Response, Compensation and Liability Act of 1980 (PL 96-510) focused on the identification of hazardous materials deposited or released into soil, ground water, surface water, and air and established a "superfund" to meet the initial costs of remediation. Cleanup costs were back-charged to generator(s). Penalties/fines from violators and taxes on chemical and petrochemical feedstocks financed the fund. Under the provisions of CERCLA, environmental monitoring is required for both newly released materials and materials leaking from abandoned and active waste respositories. Atomic spectrometry is specified for the identification and quantification of toxic elements.

The Superfund Authorization and Reauthorization Act of 1986 is a revision and extension of CERCLA.

1.4.2 Resource Recovery and Conservation Act (RCRA)

The goals of the Resource Recovery and Conservation Act of 1976 (PL 94-580) were to protect human health and environmental quality by improving solid waste management, and to conserve energy and valuable material resources. To achieve the former, the regulations require those who generate, transport, store, treat, and/or dispose of solid waste to advise the US EPA on the waste's ignitability, corrosivity, reactivity, and toxicity, and, on this basis, to identify the waste as hazardous or nonhazardous. Specific test procedures for evaluating these parameters are included in the regulations (SW-846). The toxicity evaluations are made using 100 times the original SDWA maximum contaminant levels, and heavy metal contaminants, must be determined by means of atomic spectrometry.

1.4.3 Safe Drinking Water Act (SDWA)

The goal of the Safe Drinking Water Act of 1974 (PL 93-523) is to ensure that the drinking water in the United States is safe and potable. Among the regulations established by the US EPA to meet this goal are maximum levels of and methodologies for the determination of inorganic and organic contaminants of water and some of its physical properties. Atomic spectrometry is specified for the determination of heavy metal contaminants in potable water. A unique aspect of the SDWA is the requirements that all water quality measurements be made in approved (certified) laboratories. The SDWA contains the prerequisites to certification. In 1993 the Clinton administration presented 10 recommendations improve public health protection. These were incorporated into the SDWA Amendments of 1996, which were signed into law as PL 104-182 on August 6, 1996.

1.4.4 Water Pollution Control Act (WPCA), Clean Water Act (CWA), and Water Quality Act (WQA)

While the Safe Drinking Water Act established standards for regulating the quality of potable water in the United States, the Water Pollution Control Act of 1972 (PL 92-500), the Clean Water Act of 1977 (PL 95-217), and the Water Quality Act of 1987 (PL 100-4) were enacted to control waste discharged into the nation's surface waters. Permits and effluent guidelines of the National Pollutant Discharge Elimination System (NPDES) regulate the concentrations of pollutants that may be present in wastewater before it is discharged.

The goal of the CWA was to make the nation's water fit for fishing and swimming. Two approaches were directed toward this goal: limitations on effluent discharges from industrial and domestic sources, and adoption of water quality standards by the individual states. The US EPA developed regulations for controlling effluent discharges and requiring dischargers to apply for permits from the National Pollutant Discharge Elimination System or the appropriate state system. These permits (authorized under amendments to the WPCA), issued to enforce the CWA, require, where applicable, the determination of elemental pollutants by atomic spectrometry.

1.4.5 Clean Air Act (CAA)

The goal of the Clean Air Act of 1970 (PL 95:95) was to establish and implement standards for improving the quality of ambient air to the point of eliminating all health hazards from the oxides of sulfur and nitrogen and other airborne pollutants. This act also called for the elimination of lead from automotive fuels and reductions in automotive exhaust emissions.

The Clean Air Act of 1992 (PL 101-549) mandated further reductions in airborne pollutants as well as the use of oxygenated automotive fuels during the winter months in some three dozen designated regions of the United States.

1.4.6 Occupational Safety and Health Act (OSHA)

The Occupational Safety and Health Administration (also OSHA) of the United States, with authority from the Occupational Safety and Health Act of 1970 (PL 91-596), develops health and safety standards for the workplace. Standards for numerous pollutants in the workplace atmosphere were established as were methodologies for regulatory compliance monitoring.

1.4.7 Nutritional Education and Labeling Act (NELA)

Provisions of the Nutritional Education and Labeling Act of 1992 required the inclusion of information on the concentrations of some biologically active trace elements in label statements for processed foods.

1.5 REGULATORY COMPLIANCE MONITORING

Many federal and state statutes have addressed regulatory compliance monitoring both in terms of parameters to be measured and in terms of methodologies to be employed. Atomic spectrometry is the methodology most frequently required, or at least recommended, for the determination of metals and metalloids.

2

SAMPLE COLLECTION AND PRESERVATION

Without carefully executed and fully documented procedures for sample collection, sample preservation, and sample preparation, the results obtained by atomic spectrometry are of little or no value in regulatory compliance monitoring. To relate the results obtained in the atomic spectrometry laboratory to the system under investigation, it is necessary that (1) the relationship between the sample and the system be defined, (2) the composition of the sample remain chemically and physically stable during the period of time between its collection and its analysis, and (3) the sample be converted to a solution suitable for atomic spectrometry without loss or addition of analyte.

2.1 SITE SELECTION FOR SAMPLING AIR, WATER, SOIL, SOLID AND LIQUID WASTES, AND PLANT AND ANIMAL TISSUES

The trace element composition of biological and environmental systems varies both in time and in space. These variations must be considered in the planning and execution of sampling programs and in the interpretation of laboratory results. The New Jersey Department of Environmental Protection[28] (NJ DEP) has described five ways for deciding where and when to sample:

a. Haphazard Sampling

Haphazard sampling takes the philosophy of "any sampling location will do." This situation encourages taking samples at convenient locations or times, which can lead to biased estimates of means and other site characteristics. Haphazard sampling is appropriate if the target area is completely homogeneous in the sense that the variability and the average level of the pollutant do not change over the target area. This assumption is highly suspect in most environmental pollution studies since the typical site is affected by point sources of pollution.

b. Judgement Sampling

Judgement sampling means subjective selection of sample locations by an individual. For example, judgement sampling might be used for target areas where the individual can inspect all sample locations and select those that appear to be representative of average conditions. However, the individual may select sample locations whose values are systematically too large or too small.

If the individual is sufficiently knowledgeable, judgement sampling can result in accurate estimates of site characteristics such as means and total loads even if all sample locations cannot be visually assessed. Judgement sampling can be accurate, but the degree of accuracy is difficult to quantitate.

Judgement sampling as referred to here does not refer to using prior information to plan the study, define the target area, and choose an efficient sampling design. For example, consider sampling groundwater around a hazardous waste site. The judgement and knowledge of geochemists and hydrologists is needed to make sure wells are placed in appropriate locations to detect leaks from the waste site.

c. Biased Sampling

Biased sampling is the standard for evaluating hazardous wastes sites with specific known sources. This involves collecting "worst case" samples in locations expected to be most impacted by industrial operations, such as locations of staining, drainage, and source locations, Biased sampling has been used as a means of reducing the significance of error since worst case results are generated; however, biased sampling should not be used in unknown situations since biased sampling will significantly underestimate the actual standard deviation. In summary, biased sampling is suitable when the history of a site is well known as it will determine if worst case conditions are a significant health threat.

d. Search Sampling

Search sampling is conducted to locate pollution sources or to find "hot spots" of elevated contamination. Examples include the use of gamma radiation surveys from aircraft or portable ground detectors to look for areas of elevated radiation around nuclear installations, taking cores of solid to search for buried waste, and sampling water at strategic locations in a stream network to locate a pollution source. The validity of the procedure depends on the accuracy of prior information on where or when to begin the search and on accurate measurements over time or space to guide the search. Therefore, suppose that the objective of sampling is not to estimate an average, but to determine whether "hot spots" (highly contaminated local areas) are present or to find a target of specified size. For example, it may be known or suspected that hazardous chemical wastes have been buried in a landfill, but its exact location is unknown. This section identifies methods for answering the following questions when a square, rectangular, or triangular systematic sampling grid is used in an attempt to find hot spots:

- What grid spacing is needed to hit a hot spot with specified confidence?
- For a given grid spacing, what is the probability of hitting a hot spot of specified size?
- What is the probability that a hot spot exists when no hot spots were found by sampling on a grid?

The methods referred to in this section require the following assumptions:

- The target (hot spot) is circular or elliptical. For subsurface targets this applies to the projection of the target to the surface.
- Samples are taken on a square, rectangular, or triangular grid. Calculations differ based on the grid pattern employed.
- The difference between grid points is much larger than the area sampled. Measured, or cored at grid points—that is, a very small portion of the area being studied can actually be measured.
- The definition of "hot spot" is clear and unambiguous. This definition implies that the types of measurement and the levels of contamination that constitute a hot spot are clearly defined.
- There are no measurement misclassification errors—that is, no errors are made in deciding when a hot spot has been hit.

e. Probability (Statistical) Sampling

Probability (statistical) sampling refers to the use of a specific method of random selection. Figure 2.1 shows several probability sampling designs (defined below) for spatially distributed variables.

Simple random sampling is a design where each of the sampling locations has an equal chance of being one of the sample locations selected for measurement and the selection of one location does not influence the selection of other locations. Simple random sampling is appropriate for estimating means and total loads if the target area does not contain major trends, cycles, or patterns of contamination. Note that random sampling is not equivalent to picking locations haphazardly.

In using simple random sampling, suppose a target area has a defined number of

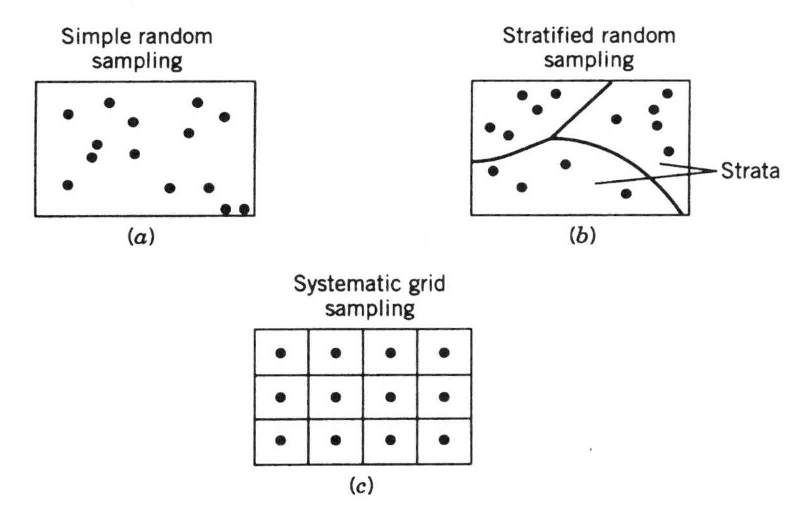

Figure 2.1 Some two-dimensional probability sampling designs for sampling over space. *Source:* Ref. 28.

sample locations(N) ; simple random sampling may then be used to select a given number (n) of these locations for measurement. The first step is to number the locations from 1 to N. Then n integers between 1 and N are drawn at random from a number table. The table consists of rows and columns of digits 0, 1, 2, . . . , 9 in random order. Once having established a starting place, then follow along horizontally or vertically in the table until the required n number are obtained. The use of such a table ensures that every unit not yet selected has an equal chance of being chosen on the next draw. The units labeled with the selected n integers are then measured.

Stratified random sampling is a useful and flexible design for estimating average pollution concentrations and total amounts. The method makes use of prior information to divide the target strata. Sampling locations are then selected from each stratum by simple random sampling. For example, a nuclear weapons test site was divided into geographical subregions (strata) and sampling locations were chosen at random within each stratum. This method provided a more precise estimate of radioactive contamination than if simple random sampling had been used without stratifying the region.

Systematic sampling consists of taking measurements at locations and/or times according to a spatial or temporal pattern; for example, at equidistant intervals along a line or on a grid pattern. Sampling designs discussed previously for the estimating of a mean or total load used a random selection procedure for choosing all sample locations to be measured. Systematic sampling is a method where only one of the locations is randomly selected. This randomly selected location established the starting place for a systematic pattern of sampling that is repeated throughout the population of N locations.

A simple example is choosing a point at random along a transect, then sampling at equidistant intervals thereafter. When sampling a geographical area, a common systematic design is to sample on a square grid. The location on one grid point is determined at random within a local region. Then the locations of the remaining grid points are completely determined by the fixed spacing between grid lines. Systematic sampling is usually easier to implement under field conditions than are simple random or stratified random sampling designs, and statistical studies indicate it may be preferred over other sampling plans for more accurate estimates of means, totals, and patterns of contamination. However, it is more difficult to estimate the variance of estimated means and other quantities than if random sampling is used. Also incorrect conclusions may be drawn if the sampling pattern corresponds to an unsuspected pattern of contamination over space and time.

In developing and executing sampling programs, inclusion of control sites is necessary for proper interpretation of laboratory results. Control sites should be located in environment and biological settings matching the system of interest but unexposed to the pollutants for which the monitoring is undertaken.

2.1.1 Site Selection for Air Monitoring

Air monitoring includes the sampling of occupational and domestic atmospheres, ambient environments, and exhaust emissions. Prevailing meteorological conditions influence the concentrations of airborne materials in ambient air and in flow streams leading to the gaseous effluent outfall. Airflow rates and patterns can lead to localized enrichments and depletions of airborne materials.

The following consideration have been found to be useful in sampling exhaust ducts and stacks. Every attempt should be made to collect samples from sections of duct showing uniform cross section because such locations are likely to demonstrate uniformity in air pressure, temperature, velocity, and composition. Locations in the vicinity of equipment tend to show variations in these parameters. Sampling sites should be located upstream from fans, pumps, and so on. Protrusions, enlargements, constrictions, and bends also lead to nonuniform conditions. Sampling sites should be located downstream from such irregularities.

The ambient environment should be sampled at sites away from commercial and industrial exhaust gas outfalls. The sampling sites should be selected to avoid co-collecting of materials carried by rainwater or snowmelt. Protective housing are recommended for long-term sampling of the ambient environment. The guidelines of the American Society for Testing and Materials (ASTM) suggest sampling through vertical inlets to such housing located 2.5–5 m above ground level whenever possible.[29]

Sampling sites for the domestic and occupational atmosphere should be located away from heating and cooling vents. Locations adjacent to office machines and factory equipment should be avoided unless these are the focus of the monitoring. Protective housings are recommended when the potential for accidental or deliberate contamination exists.

Personal samplers are required when the goal of the monitoring is the assessment of occupational/public health. Breathing-zone air should be sampled for this purpose.

2.1.2 Site Selection for Potable Water Monitoring

The best site for sampling potable water is the point of consumption. The Safe Drinking Water Act[30] specified potable water sampling as follows:

1. Ground water systems shall take a minimum of one sample at every point to the distribution system which is representative of each well after treatment (hereafter called a sampling point) beginning in the compliance period starting January 1, 1993. The system shall take each sample at the same sampling point unless conditions make another sampling point more representative of each source or treatment plant.
2. Surface water systems shall take a minimum of one sample at every point to the distribution system after any application of treatment or in the distribution system at a point which is representative of each source after treatment (hereafter called a sampling point) beginning in the compliance period starting January 1, 1993. The system shall take each sample at the same sampling point unless conditions make another sampling point more representative of each source or treatment plant.
3. If a system draws water from more than one source and the sources are combined before distribution, the system must sample at an entry point to the distribution system during periods of normal operating conditions (i.e., when water representative of all sources is being used).

The importance of obtaining a representative sample cannot be over-emphasized. To this end, it is necessary to identify multiple sampling points within the system.

2.1.3 Site Selection for Ambient Surface Water Monitoring

As is the case with potable water monitoring, the need to obtain representative samples is a major factor in selecting sites for surface water monitoring. Water from the swift flow of a stream or river channel will tend to show higher trace element concentrations than the more placid near-shore water, since the running water will be higher in suspended solids. During periods of rainfall or snowmelt, however, the near-shore water may show higher trace element concentrations as a result of contributions from surface runoff. For these reasons, the establishment of multiple sampling points along a bank-to-bank transect, with distances between sampling points corresponding to the greater of 10% of the transect or 10 m is recommended. Seasonal variations may necessitate repetitive sampling quarterly or even monthly.

Stratification of lakes deeper than 5 m often leads to variations in the chemical and physical parameters of samples collected at different depths.

Stratification is also a consideration in ocean sampling. Various chemical species may be stratified at different depths. In addition, the composition of near-shore water is often different from deep ocean water. Estuarian sampling is even more complex because strata move up and down rivers unevenly.

Large bodies of water usually require the establishment of a multidimensional grid for locating sampling points and times. Such a grid is four-dimensional: shore-to-shore, upstream-downstream, and surface-to-bottom. Time is the fourth dimension. Financial considerations often limit the width, length, depth, and frequency of the sampling points. The duration of sampling should be 10 times longer than the longest period of interest, but this too may become impractical in terms of costs.

In establishing sampling points for ambient surface water monitoring, the US EPA[31] recommends avoiding location vulnerable to airborne contamination of the samples. Among the possible sources of contamination cited were automobile exhaust; cigarette smoke; corroded or rusted bridges, pipes, poles, or wires; nearby roads; and areas in which bare soil was subject to wind erosion.

The New Jersey Department of Environmental Protection has made similar recommendations and added special considerations of tidal stages and currents in establishing sampling schedules.[28] In addition to recommending the four-dimensional sampling grid, the NJ DEP recommends approaching the sampling site from downstream and, if a boat is used, collecting the sample as far from the stern as possible.

2.1.4 Site Selection for Ambient Groundwater Monitoring

The vulnerability of groundwater to contamination is affected by depth, recharge rate, soil composition, and topology as well as by the persistence or mobility of the contaminant. The New Jersey Department of Environmental Protection recommends specifying the objectives of groundwater sampling programs prior to their implementation.[28] Some possible objectives are:

- determining the potability of a private or municipal supply well,
- establishing the presence or absence of contamination in a given study area,

- assessing the vertical and horizontal extent of a contamination plume,
- measuring the direction of ground water flow,
- evaluating the hydrologic characteristics of an aquifer, and
- identifying potential receptors and receptor pathways.

The design of a groundwater sampling program is dependent on the presence, locations, pumping schedules and rates, and construction of monitoring wells. In cases of newly suspected groundwater contamination, strategically placed monitoring wells may not be available. In such cases, the New Jersey Department of Environmental Protection suggests with reservation that groundwater samples from private wells be used only to make preliminary assessments on the presence or absence of contamination. The department maintains, "Private supply wells should only be used to evaluate potability and to gather preliminary data. These wells should not be used as a substitute for properly located and constructed monitor wells."[28]

The multidimensional grid model for establishing surface water sampling points is applicable to locating groundwater monitoring wells. The New Jersey Department of Environmental Protection has developed specifications for the construction of groundwater monitoring wells in bedrock, unconsolidated, and confined aquifers.[1]

When constructing monitoring wells, care must be taken during the drilling process not to cross-contaminate aquifers with loosened topsoil possibly laden with agricultural and industrial chemicals. Material blanks are important because well construction and the materials used can profoundly influence the chemical composition of samples.[32]

2.1.5 Site Selection for Sampling Soil

The multidimensional grid model for establishing surface water sampling points is applicable to soil sampling. Both surface and subsurface sampling are important for determining contamination in the soil and its potential for polluting groundwater. Samples collected from the uppermost 60 cm are classified "surface," while those collected from depths greater than 60 cm are classified as "subsurface." Rump and Krist[33] recommend 20 sites for each 10,000 m^2 of surface grid. Crépin and Johnson[34] propose 5–10 sampling sites for test areas smaller than 0.5 ha (1 ha = 10,000 m^2). The New York State Department of Environmental Conservation[35] recommends spacing grid lines at 10-foot intervals, conducting preliminary sampling at every tenth intersection, and then collecting samples from every intersection in regions of the site that require more intense evaluation.

2.1.6 Site Selection for Sampling Liquid Wastes

Industrial and municipal effluent outfalls present fewer problems in the selection of sample sites than do rivers and streams. The same considerations of flow patterns and storm water inputs must be made, but on a smaller scale. One or two sites per effluent stream usually are sufficient. These should be located near the source of the effluent; the second site, when established, should be near the confluence of the effluent stream and the receiving body of water. The effluent streams are more likely to show variations on a time basis than on a site-to-site basis.

Sampling site selection for waste lagoons, settling ponds, and so on parallels that for lakes. The number and depths of these sites will depend on the system being sampled. Unlike the effluent streams, these lagoons and ponds are more likely to show horizontal and vertical site-to-site variations. Meteorological conditions may also give rise to variations in composition on a temporal basis.

Sites selected for monitoring wells to collect landfill leachate must reflect consideration of the direction of groundwater flow and the depth(s) of the adjacent aquifer(s). In addition to the size of the landfill, the geological, hydrological, and meteorological characteristics of the site must be carefully evaluated. Number and location(s) of monitoring wells usually are determined on the basis of the unique features of each and every landfill.

Liquid wastes contained in drums and barrels, as well as those contained in storage tanks and rail or highway tankers, are sampled on a container-by-container basis with awareness that the contents of the container may be multiphasic or stratified. Each container should be treated as a separate system unless the contents are known to be the same.

2.1.7 Site Selection for Sampling Sediments and Sludges

Sites for sampling marine, estuarian, and aquatic sediments as well as the sediments in waste lagoons and settling ponds may be selected by means of the multidimensional grid model used in establishing soil sampling points.

2.1.8 Site Selection for Sampling Solid Wastes

The identification of specific sampling sites may or may not be necessary for some solid wastes. Grab samples of solid wastes such as dried sewage sludge or incinerator ash can be obtained by the conventional coning and quartering methods.[36–38] The following sampling points for the collection of solid sludge were recommended in the Sampling Procedures and Protocols for the National Sewage Sludge Survey[39]: belt filter press, centrifuge, vacuum filter, sludge press, drying beds, and compost piles. The NJ DEP recommends sampling the nonweathered surface of drying beds at 5-foot intervals.[1] Other kinds of solid waste require consideration on a case-by-case basis.

The federal Environmental Protection Agency[40] has identified six types of waste based on the uniformity of the generating processes and on the degree of homogeneity with which the contaminants are distributed within the waste. These six types of waste are:

- Uniformly homogeneous
- Nonuniformly homogeneous

- Uniformly, randomly heterogeneous
- Nonuniformly, randomly heterogeneous
- Uniformly, nonrandomly heterogeneous
- Nonuniformly, nonrandomly heterogeneous

In general, the more homogeneous the waste, the fewer sampling sites need be identified. The number of samples needed to attain a given level of precision in their arithmetic mean n depends on the concentration range of contaminant in the waste r [$s^2 = (r/4)^2$], and the acceptable uncertainty in the mean estimation of the concentration of the contaminant d:

$$n = \frac{t^2 s^2}{d^2}$$

The value of t depends upon the value of n. Table 2.1 gives some values for t at several levels of precision.

At 90% precision, nearly 300 samples of waste containing 50–450 ppm chromium contaminant would be required to achieve an uncertainty in their mean value of ±10 ppm. In a waste with a more homogeneous distribution of chromium, from 50 to 150 ppm, fewer than 20 samples would be required. The corresponding numbers of samples for 80% precision are approximately 150 and 10.

Table 2.2 lists several conditions in which solid and liquid wastes may be found and the corresponding sampling points.

2.1.9 Site Selection for Sampling Plant and Animal Tissues

The requirements for selecting sites to collect plant and animal tissues depend, for the most part, on the objectives of the investigation. The establishment of multidimensional grids can be useful in selecting sites for the collection of plant and animal specimens in the test and control areas to evaluate the levels of selenium in pasture grass-

TABLE 2.1 *t* Values

	Precision Level			
n	80%	90%	95%	99%
5	1.476	2.015	2.571	3.365
10	1.372	1.812	2.228	2.764
15	1.341	1.753	2.131	2.602
20	1.325	1.725	2.086	2.508
25	1.316	1.708	2.060	2.485
30	1.310	1.692	2.042	2.452
40	1.303	1.684	2.021	2.423
60	1.269	1.671	2.000	2.390

TABLE 2.2 Recommended Sampling Points for Solid Wastes

Waste Containment	Sampling Point[a]
Drum	Drum opening
Rail/highway tanker	Each hatch
Pond, pit, lagoon	Center of each horizontal grid square at surface, middepth, and bottom
Storage tank	Each hatch
Barrel, fiber drum, sack	Container center and two points diagonally opposite initial point of entry
Waste pile	Top of pile and two or more points diagonally opposite top of pile
Soil	Center of each cube in the three-dimensional grid

[a]The number of grid points is determined by the desired number of samples to be collected, which when combined should give a representative sample of the waste. See text.

es or to assess that concentrations of mercury in shellfish. On a time grid, pre- and postshift and/or work- and rest-day urine collections from factory and office workers at a zinc smelter may be useful for determining cadmium intoxication. Gross sampling of plant and animal tissues is adequate for identifying environmental exposures. A more detailed and more controlled sampling program is needed to investigate the mechanisms of biochemical detoxification or mineral metabolism. Regardless of the objective, a statistically significant number of specimens is required.

2.2 DISCRETE VERSUS COMPOSITE SAMPLING

The mode and frequency of sampling depends both on the homogeneity of the system being sampled and on the information being sought from the system. A discrete or grab sample consists of a single specimen. It reflects the characteristics of the system at the point in the time–space continuum corresponding to its collection. Collection may be either manual of automatic. A composite sample, on the other hand, is a mixture of several grab samples, collected either from a single site at various time intervals or from various sites during the same time interval. The amount of an individual sample added to the total mixture should reflect its contribution to the system under investigation (flow rate an effluent, population of a botanical species in a forest, etc.). Composite samples tend to average localized irregularities in the system. Like grab samples, composite samples can be collected either manually or automatically. Discrete samples are preferred over composite samples when (1) the system to be sampled is not continuous, such as intermittent discharges from several holding tanks, (2) the system is known to show relatively constant characteristics, or (3) it is desired to determine whether the system shows extremes in composition. In general, the analysis of a large number of samples collected at different times from different sites in the system provides much more information than could possibly be obtained from the analysis of a single composite sample. (The victims of Minamata disease find little comfort in learning that the average mercury concentration in sea fish is below 0.5 ppm.) An advantage of composite sampling is the reduction in laboratory workload and the corresponding reduction in monitoring costs.

2.2.1 Air Samples

Collecting samples of airborne materials for elemental analysis frequently takes place over time intervals ranging from hours[41] to months.[42] The development of high volume air samplers was responsible for time reductions to the lesser value. Even samples collected over prolonged periods of time are usually considered to be grab samples from a specified site.[43]

2.2.2 Water Samples

The presence or absence of contaminants in water and wastewater is usually evaluated from grab samples taken from multiple sites in the system on a routine basis, (e.g., annually, quarterly). The number of sites is determined by the size of the system. The number of sites depends on the variability of sample composition.

2.2.3 Soil Samples

The presence or absence of contaminants in soil is usually evaluated from grab samples taken from surface and subsurface sites in a multidimensional grid.

2.2.4 Liquid Wastes

Grab samples are collected from liquid waste ponds, pits, and lagoons at sites established with a multidimension grid. Grab samples of liquid wastes are also collected from drums and tanks.

2.2.5 Solid Wastes

Solid waste samples, depending on the origin of the waste and on the information sought, are usually collected as composites. Such wastes should be homogeneous. When homogeneity is lacking, grab sampling becomes necessary.

2.2.6 Plant and Animal Tissue Samples

Both grab sampling and composite sampling have been used to collect specimens of plant and animal tissues. Grain samples for the determination of arsenic, shellfish and finfish samples for the determination of mercury, and milk-free infant formula samples for the determination of chromium were composited prior to the analyses. Grab sampling is used to collect blood for the determination of lead. Composite sampling is used to collect material for elemental analysis in food processing and other production line operations. Grab sampling is used in collecting materials of clinical or laboratory significance.

2.3 SAMPLE CONTAINERS

The function of the sample container is to hold the sample from the time of collection to the time of analysis. During this period, the container itself may neither add to nor remove from the sample. In addition, the container must protect the sample from contamination and/or loss during this period.

2.3.1 Containers for Air Samples

With few exceptions, samples of airborne materials are collected by drawing measured volumes of air through cellulose acetate–nitrocellulose filters or through glass fiber filters. The loaded filters from "Hi-Vol" samplers are folded in half, carefully enclosing the collected materials, and the folded filters are placed in clean paper or plastic envelopes for return to the laboratory.[44,45] The loaded filters from personal samplers are sealed in their original cassettes and returned to the laboratory for analysis.[46] Liquid-filled impingers have been specified for the collection of air samples for hexavalent chromium analysis.[47] The impinger serve as containers for returning the samples to the laboratory. Hopcalite-filled sorbent tubes are required for the containment of mercury collected from workplace atmospheres.[48] Filter, impinger, and sorbent tube field blanks must be included in the monitoring protocol to allow the establishment of background values for the collection media. Typical background values for some commercially available cellulose ester and glass fiber filters are listed in Table 2.3.

TABLE 2.3 Typical Purity of Cellulose Ester (CE) and Glass Fiber (GF) Filters[a]

Element	CE (ppm)	GF (ppm)	Element	CE (ppm	GF (ppm)
Al	2	mc	Mg	1.8	mc
Ag	nd	<0.2	Mn	<0.05	1.2
B	<2	mc	Mo	nd	0.5
Ba	<1	mc	Na	40	mc
Be	nd	0.3	Ni	nd	2.5
Bi	nd	1.0	Pb	0.2	2.5
Ca	13	mc	S	<5	nd
Cd	nd	Sb	<0.02	nd	
Co	nd	0.3	Si	<2	mc
Cr	0.3	1.2	Sn	nd	2.5
Cu	0.4	0.7	Ti	nd	2.5
Fe	6	25	V	nd	2.5
K	1.5	mc	Zn	0.6	25

[a]mc, major component; nd, not detected.

Berg et al.[49] measured trace elements in unused air filters of 18 different types obtained from four different suppliers. Glass fiber filters showed the highest blank values, and filters made from poly (tetrafluoroethylene) (PTFE, Teflon) had the lowest levels of trace elements.

2.3.2 Containers for Liquid Samples

The regulatory and advisory agencies have identified containers suitable for water and wastewater samples. The New Jersey Department of Environmental Protection specifies,[1] and the United States Environmental Protection Agency,[50] the American Society for Testing and Materials,[51] and the American Public Health Association[52] recommend, chemically resistant glass or chemically resistant plastic bottles and closures with inert liners. Moody and Lindstrom[53] have evaluated several plastic containers for trace element samples, and Laxen and Harrision[54] and Kammin et al.[55] have critically reviewed cleaning protocols for freshwater sample containers.

2.3.2.1 Cleaning of Liquid Sample Containers The following procedure has been recommended for cleaning sample containers prior to use:

1. Wash the bottles and closures with a good quality detergent in hot water.
2. Rinse the bottles and closures well with tap water.
3. Rinse the bottles and closures with 1:1 nitric acid.
4. Rinse the bottles and closures well with tap water.
5. Rinse the bottles and closures with successive portions of high purity water.
6. Invert the bottles and closures in a dust-free environment for drying.[56]

In addition to the method just described, the US EPA more recently recommended, "Laboratory glassware, including the sample collection cubitainer and the polyethylene sample storage bottle, as well as the filtering apparatus should be thoroughly washed with (1+1) nitric acid, tap water, (1+1) hydrochloric acid, tap water and finally distilled water in that order."[57] Another US EPA method states: "All reusable labware. . . , including the sample container, should be cleaned prior to use. Labware should be soaked overnight and thoroughly washed with laboratory-grade detergent and water, rinsed with water, and soaked for four hours in a mixture of dilute nitric and hydrochloric acid (1+2+9), followed by rinsing with water, ASTM type I water and oven drying."[58] Precleaned sample containers are available from a variety of vendors, including Eagle-Picher (Miami, OH), Qorpac (Bridgeville, PA), and Wheaton (Millville, NJ). Specifications for Wheaton Clean-Pak containers are compared to the US EPA guidelines[59] in Table 2.4. Field blanks and controls must be included in the monitoring protocol to validate the absence of contamination and loss.

2.3.3 Containers for Solid Samples

Samples of soils and solid wastes are conveniently collected in wide-mouthed glass or plastic jars that have been cleaned by the procedures just outlined. One-liter, wide

TABLE 2.4 Practical Quantification Limits for Clean-Pak Field Blanks

Element	Wheaton Clean-Pak (μg/L)	US EPA Recommendation (μg/L)
Aluminum	<80	100
Antimony	<5	5
Arsenic	<2	2
Barium	<20	20
Beryllium	>0.5	1
Cadmium	<1	1
Calcium	<500	500
Chromium	<10	10
Cobalt	<10	10
Copper	<10	10
Iron	<50	500
Lead	<2	2
Magnesium	<100	500
Manganese	<10	10
Mercury	<0.2	0.2
Nickel	<20	20
Potassium	<750	750
Selenium	<2	3
Silver	<5	10
Sodium	<500	500
Thallium	<5	10
Vanadium	<10	10
Zinc	<10	20

mouthed polyethylene jars with linerless polyethylene lids were used to collect samples of sediments and soils.[60] The jars were washed with detergent, thoroughly rinsed, and washed in 1:1 nitric acid and then in 1:1 hot hydrochloric acid. Each wash was followed by thorough rinsing with deionized, distilled water. Jars serving as shipping blanks were sent into the field, returned empty, and rinsed with nitric acid. Analysis of the nitric acid rinses showed nondetectable levels of aluminum, antimony, barium, beryllium, bismuth, cadmium, cobalt, copper, lead, molybdenum, nickel, selenium, silver, strontium, tin, titanium, and vanadium. The following were found in microgram amounts: calcium, 117; chromium, 16; iron, 108; magnesium, 7; manganese, 3; potassium, 10; sodium, 36; zinc, 15. If the jars had been filled to capacity, and if all the acid-leached elements contaminated the samples, the effect of the added calcium and added manganese would have been ~ 0.1 ppm and ~ 3 ppb, respectively.

2.3.4 Containers for Plant and Animal Tissue Samples

Wide mouthed plastic and glass containers have been used for plant and animal tissues. Blood samples are routinely collected in evacuated glass tubes. Nackowski et al.[61] have made a survey of trace metal contaminants in the blood collection tubes produced by several manufacturers. Their study indicates that zinc, lead, and cadmium contamination of the sample can be a significant problem during normal handling, shipping, and storage. Lecomite et al.[62] found significant zinc contamination of blood samples collected with evacuated tubes from one particular manufacturer. Handy[63] identified the rubber stoppers of these tubes as the major source of the zinc contamination, but Versieck et al.[64] found that the amounts of zinc (and of copper) transferred from stoppers to samples was insignificant compared to the natural levels of these elements in blood serum. Koirtyohann and Hopps[65] have ranked fluorinated hydrocarbon containers as most suitable, polypropylene or polyethylene containers as acceptable, and containers made from other plastics or glass as less satisfactory for specimens of human tissues. Katz,[66] who reviewed the collection and preparation of biological tissues and fluids for trace element analysis concluded:

> There are many potential sources of error in collecting and preparing biological tissues and fluids for the measurement of their trace element contents. It is possible, however, to obtain reliable results when these sources of error are known and special precautions are incorporated into the analytical procedures. A rigorous program of quality control/quality assurance is needed to confirm the reliability of the results obtained for trace element concentrations in biological tissues and fluids.

2.4 SAMPLE COLLECTION

Sample collection is a crucial part of compliance monitoring. Special efforts must be taken to assure that the material evaluated in the laboratory is representative of the system under investigation. To permit correlation of laboratory results with the sampling site, samples must be properly identified at the time of collection. Sample identification is made both from information on the sample label and from information in the sample or field logbook. Information must be recorded at the time the sample is collected. Label information should include the sample identification number, the place of collection, the time and date of collection, and the name of the sample collector. The information recorded in the sample or field logbook includes, where appropriate, the following: name of site; date and time of site entry; name(s) and address(es) of field investigator(s) and sample collector(s); purpose of sampling; type of process/activity that produced the (suspected) contamination; date(s) and time(s) of sample collection and sample identification number(s); description(s) and location(s) of sampling point(s); weather conditions; field measurements of pH, conductivity, temperature, flammability, explosiveness, and so on; water level measurement(s) and depth(s) to bottom(s) of monitoring well(s); presence/absence of immiscible liquids in monitoring well(s); volume(s) of water purged and time(s) to recharge well(s); physical description(s) of sample(s); number(s) and size(s) of sample(s) collected; procedure(s) and frequency(ies) of sampling equipment decontamination; time(s), date(s), and conditions for transport of samples to laboratory; and other relevant information.

2.4.1 Air Sampling

Adequate sampling of atmosphere is often difficult because particulates suspended in the air are not uniformly dispersed. They do not follow the airstream lines in the atmosphere, and they are frequently subjected to forces other than aerodynamic ones. Before an analysis of samples is performed for purposes of environmental monitoring, careful sampling must be conducted.

Adequate sampling of particulate matter involves, in most cases, isokinetic sampling (particles entering the sampling orifice without undergoing a change in velocity). Anisokinetic sampling leads to the collection of a disproportionately small number of the larger particles. Such sampling errors are minimal in the open atmosphere. When samples are collected from stacks or ducts, the errors associated with anisokinetics should be considered. The concentrations of some trace elements are enriched on the smaller particles.[67] Analysis of a sample collected anisokinetically from a stack would show erroneously high values for these trace elements in the particulate emissions.

The sampling of airborne particulates is most frequently carried out by filtration. For subsequent chemical analysis, readily soluble filter media are frequently selected. Sampling from high temperature environments requires the use of alumina or quartz fiber filters. Some typical filter recommendations are listed in Table 2.5.

The use of inertial collection devices, which retain particles in specific size ranges, adds a dimension of selectivity. While the single-stage impactor (wet impinger) is suitable for collecting airborne particulates from a variety of atmospheric environments, the multistage (cascade) impactor retains the larger particles in the earlier stages, thereby permitting simultaneous collection and size sorting.

TABLE 2.5 Recommendations for Air Sampling

Analyte	Filter[a]	NIOSH Reference[b]
Aluminum	0.8 μm MCE	7013
Arsenic	0.8 μm MCE	7900
Barium	0.8 μm MCE	7056
Beryllium	0.8 μm MCE	7102
Cadmium	0.8 μm MCE	7048
Calcium	0.8 μm MCE	7020
Chromium	0.8 μm MCE	7024
Cobalt	0.8 μm MCE	7027
Copper	0.8 μm MCE	7029
Lead	0.8 μm MCE	7082
Mercury	Hopcalite tube	6009
Tungsten	0.8 μm MCE	7074
Vanadium oxides	Cyclone + 5 μm PVC	7504
Zinc	0.8 μm MCE	7030

[a]MCE, mixed cellulose esters; PVC, polyvinyl chloride.
[b]*NIOSH Manual of Analytical Methods*, 4th ed., Cincinnati, OH, 1994.

2.4.2 Potable Water Sampling

Sampling potable water supplies may involve the collection of water samples that are representative of untreated water (raw water), treated water (plant-delivered water), and water from the distribution system (system water).[28] Raw water samples can be collected from the pumps used to transport it to the treatment facility, and plant-delivered water can be sampled as it leaves the treatment facility. Two types of sample can be collected from the distribution system: first-draw and flushed. A first-draw sample is water that immediately comes out when the tap is first opened. This type of sample is useful when evaluating whether plumbing materials are contributing lead or other contaminants to the water supply. Flushed samples, collected after piping has been evacuated, should be representative of the water flowing in public water systems.

When collecting system samples, the water should be allowed to run from the spigot to waste until a volume equal to that contained in the service line has flushed through the collection point. For 50 feet of 0.75- inch service line, flushing for 5 minutes in adequate time to replace twice the volume contained in that length of line. Water softeners, filters, aerators, and so on must be removed prior to flushing and sampling.

Glass or plastic containers prepared as described earlier are filled with samples of raw, plant-delivered, or system water.

2.4.3 Groundwater Sampling

When samples are collected from municipal wells, industrial wells, domestic wells, or monitoring wells, flushing or pumping of the well is almost always required to ensure a representative sample of the groundwater.[1,8,68] During evacuation, the pump intake should be maintained at a depth 6 feet below the surface of the water standing in the well casing, and the rate of pumping should be low enough to avoid overpumping or pumping to dryness. The sample should be collected no more than 2 hours after 3–5 volumes of water have been purged from the well. The approximate volume of water in gallons V_g per foot of depth in a well casing having diameter d inches is:

d (in.)	V_g/ft
2	0.16
4	0.65
6	1.5
8	2.6
10	4.1
12	5.9

Devices needed for pumping or collecting water from wells include portable centrifugal pumps, 3- or 4-inch submersible pumps, bailers, or Kemmerer-type samplers. After the well has been flushed, the samples are collected in appropriate containers. Field blanks should be included as a part of the quality control/quality assurance program.

2.4.4 Surface Water Sampling

Collection of samples from streams, rivers, ponds, lakes, and oceans is accomplished with a variety of devices, including the Wheaton dip sampler, the Kemmerer depth sampler, and the Bacon bomb sampler.[28] The selection of sampler depends on factors such as the width, depth, and flow at the sampling point; whether the sampling point is onshore or offshore; and the information sought from the monitoring. Some of the advantages and disadvantages associated with using these samplers are summarized in Table 2.6

It is frequently possible to collect surface water samples directly into the sample bottle from onshore by using a sampler with extension capabilities. When collecting samples from bodies of water, shallow enough to permit wading, the sample bottle should be immersed mouth down to middepth and inverted. To avoid the collection of disturbed sediment, the sampling point should be approached from downstream.

Offshore sample collection often requires the use of the devices just cited. Offshore surface water samples should be collected from as far forward as possible to minimize the potential for contamination by metallic emissions in engine exhaust and in engine wear particles. For the same reason, the sampling point should be approached from the downstream direction. Sometimes it is possible to collect offshore surface water samples directly into the sample bottles. Samplers must be used to access subsurface sampling points. Subsurface samples must be transferred to sample bottles, and the sampler must be decontaminated prior to reuse.

TABLE 2.6 Comparison of Water Samplers

Sampler	Container	Advantages	Disadvantages
Wheaton dip sampler	Glass	Ease of operation Collection bottle may serve as sample container	Sampling depth limited by length of sampler Brakage of bottle
Kemmerer depth sampler	PVC	Ability to sample at discrete depths Ability to sample at great depths	Collection tube is exposed to overlying media Requires transfer from collection tube to sample bottle
Bacon bomb sampler	Brass or stainless steel	Ability to sample at discrete depths Ability to avoid exposure to overlying media during descent and ascent	Difficult to clean Requires transfer from colelction bomb to sample bottle May contaminate sample with Cu, Zn

More important than the technical aspects of sample collection are those of water safety. Approved flotation devices should be worn during all phases of onshore and offshore sampling of surface waters, as well as during all phases of wastewater sampling, as described in the next section.

2.4.5 Wastewater Sampling

Municipal and industrial wastewater may be collected as either grab or composite samples depending on the data quality objectives of the monitoring program. When the monitoring program is required for compliance to state or federal regulations, the sample type (s) and location (s) for collection are often specified by an agency of the government.[69] In the United States, these may be indicated on the National Pollutant Discharge Elimination System permit issued by the federal Environmental Protection Agency, or on the permits issued by primacy states. Operating personnel familiar with the facility should be available to identify influent and effluent streams, equalization tanks, and so on. Samples should be taken from below the surface of the stream, lagoon, or pond, to avoid contamination with floating debris.

Whenever possible, the sample should be collected directly into the sample container. This may be accomplished with devices such as the Grab Sampler I, II, or III (Wheaton, Millville, NJ). The Grab Sampler II will allow collection of wastewater directly into 1-liter sample bottles at depths as great as 5.5 m.

Composite sampling, often used for monitoring municipal or industrial wastewater discharges, provides a sample that is representative of the discharge over a defined period of time (e.g., 24 hours). Composite samples can manually collected, hourly increments over a 24-hour period, or automatic composite samples may be used at significant cost savings in long-term monitoring programs.

2.4.6 Sampling Containerized Liquid Wastes

The composite liquid waste sampler (COLIWASA) permits collection of representative samples from multiphase liquid wastes and low density sludges having a wide range of viscosity's and solids contents. The device is available from vendors such as Wheaton, which can supply a model for use at depths of 3–4 m, and NASCO (Fort Atkinson, WI). Typically the COLIWASA collects 150 mL of sample for each 300 cm of immersion in liquid waste. The design of the COLIWASA allows collection of sample(s) without displacing the overlying liquid. The assembly is lowered into the drum, tank, or pond with the stopper rod depressed, permitting the liquid to flow in through the open bottom. When the unit is filled, the bottom is closed by pulling up on the stopper rod. The sample is drained from the COLIWASA into an appropriate container by depressing the stopper rod.

2.4.7 Sampling Soils, Sediments, and Sludges

Equipment for collecting surface samples (those within the upper 60 cm) of soils, sediments, and sludges includes scoops and trowels, sampling tubes, augers, coring devices, and dredges.

The Shelby tube sampler has thin walls and a tapered cutting head that facilitates penetration into the soil to be sampled. It is inserted into the soil, sediment, or sludge without twisting or excessive force, allowed to remain in place for a few minutes, and then rotated through at least two full revolutions to shear off the sample column at its bottom. The tube is withdrawn, and the sample can be extruded into an appropriate container. Alternatively, the ends of the tube can be sealed, and it can serve as the sample container.

The split spoon sampler is a length of carbon or stainless steel, split longitudinally and fitted with a drive shoe and head. The sampler is driven into the ground with a sledgehammer, and then withdrawn and opened. Insertion into sediments and sludges requires less pressure. The sample can be split lengthwise with a clean stainless steel spatula, and the splits can be transferred to appropriate containers.

The bucket auger consists of a strong central shaft and a sharp spiral metal blade. When the tee handle on the central shaft is rotated in the clockwise direction, the blade cuts into the soil. As the blade moves downward, the loosened soil is discharged upward, where upon it is collected in a second auger. The soil collected in this manner is removed periodically and transferred to labeled sample jars. Sampling depths of nearly a meter can be reached with the bucket auger.

The silver bullet sampler is a coring device consisting of a brass cylinder fitted with a borosilicate glass inner sleeve, a serrated cutting ring, and a tee handle for turning the corer into the soil, sediment, or sludge. Cores are limited to 50 cm with the silver bullet sampler. These are obtained by pressing the sampler into the ground while rotating the tee handle clockwise. When the desired depth is reached, the sampler is rotated counterclockwise through at least two full revolutions to shear off the sample column at its base. Then the sampler is withdrawn. The borosilicate glass inner sleeve containing the core can be capped to become the sample container, or the core can be transferred to an appropriate container.

The Veihmever sampler is a coring device suited for use to depths of 3 m in sediments and sludges as well as in stone-free soil. The sampler is driven to the desired depth into the soil, sediment, or sludge with the drive hammer, rotated through at least two revolutions to shear off the sample column at its base, and withdrawn manually or with the aid of the puller jack and grip. The sample column is recovered by inverting the Veihmever sampler and gently tapping the drive head against the drive hammer and transferred the sample to an appropriate container. Special precautions are needed when samples are collected from multiple sites because the Veihmever sampler is particularly difficult to decontaminate.

The Ponar dredge is a clamshell–type scoop able to collect samples from sediments and sludges ranging from fine silts to granular particulates. The petite version of the Ponar dredge is light enough to be used without a winch or crane. It has a sampling area of 232 cm^2, but its penetration is only a few centimeters. Unlike coring devices, augers, and tube samplers, the Ponar dredge is not capable of collecting undisturbed samples. The Ekman dredge and the Smith–McIntyre dredge produce less disturbance when the bottom is contacted. These are compared in Table 2.7.

The Sludge Getter (Wheaton) is useful for the collection of liter samples as depths up to 1.5 meters. Its conical bottom and 15 kg mass enhance penetration into

TABLE 2.7 Comparison of Bottom Samples for Sediments and Sludges

Device	Characteristics
Ponar dredge	Safe, easy to use End plates minimize escape of sample Produces shock wave Can become buried in soft bottom material
Ekman dredge	Useful in soft bottom materials and calm water Collects reproducible, standard sample volume Reduced shock wave Not useful on bottoms covered with vegetation Not useful in rough water
Tall Ekman dredge	Useful in soft bottom materials and calm water Collects reproducible, standard sample volume Reduced shock wave Does not lose sample over the top Not useful on bottoms covered with vegetation Not useful in rough water
Peterson dredge	Sturdy and simple contruction Well suited for hard bottoms and for fine silt May lose some sample Sample volume is not constant
Smith–McIntyre dredge	Useful in rough water Flange on jaw reduces sample loss Screen reduces shock wave Useful for both fine and granular samples Large, complicated, and heavy

bottom materials. After penetration, the Sludge Getter is opened and reclosed from the surface to collect and retain the sample. Its fabrication from 316 stainless steel simplifies decontamination, but the possibility of contamination with nickel and chromium is a cause for concern.

The Sludge Judge (NASCO) allows sample collections at depths approaching 4.5 m. A 1 cm immersion into the bottom material results in the collection of approximately 3 mL of sample. This is accomplished by slowly lowering the Sludge Judge into the sampling point, opening and reclosing the check valves, retrieving the Sludge Judge, and emptying its contents into an appropriate sample container.

Sampling procedures and protocols for sewage sludges have been described by the US EPA.[70]

2.4.8 Sampling Solid Wastes

Specific procedures for sampling solid wastes are described in the *Federal Register.*[71] Kratochvil and Taylor[72] have presented an excellent overview on the fundamentals of sampling solid waste, and the Bendix Corporation[73] has developed an audiovisual training program entitled "Sampling for Toxic Substances." Procedures for sampling solid waste have been the subjects of several reports at ASTM symposia on solid waste testing.[74–78]

A variety of sampling tubes have been used to collect samples from containerized solid wastes, waste piles, and sludge drying beds. The sampling tubes are inserted horizontally when samples are collected from solid waste piles and from bags, drums, and barrels containing powdered or granular solid wastes. Samples collected from drying beds and from shallow containers or soil surface containing dry waste are often taken as vertical cores. Samples collected from drying beds and from shallow containers or soil surface containing dry waste can also be collected with a clean trowel or scoop.

Triers are used to withdraw sample cores of loose solid material from waste piles, bags, hoppers, and so on. Cores are most readily obtained from slightly moist or compacted material because then the core, which is cut by rotating the trier, will remain consolidated when the trier is removed from the waste. Triers are used only when horizontal insertion into the waste is possible. Triers are often fabricated from stainless steel, and they are available in lengths ranging from 50 to 100 cm.

Thiefs consist of two concentric tubes with evenly spaced openings along their lengths. A thief is inserted horizontally into the waste, with its openings opposed to each other. After insertion, the openings are aligned by rotating the inner tube, thus allowing inflow of the sample. When the inner tube is filled, it is rotated to the closed position, and the thief is withdrawn. Like the trier, the thief often fabricated from stainless steel. Chromium-plated tube samplers should be avoided. Waste cores collected with tube samplers should be promptly transferred to sample containers.

Solid wastes deposited horizontally in sludge drying beds or shallow containers or surface soil containing dry waste are sampled as vertical cores using essentially the same procedures as those described for soils in Section 2.2.3.

2.4.9 Sampling Biological Tissues

According to principles developed by the National Institute of Occupational Safety and Health (NIOSH)[79] for the collection of biological samples that reliable laboratory data cannot be obtained unless:

1. The correct specimen is collected.
2. The specimen is collected at the right time after exposure to the toxic substance.
3. The proper container, with preservative or anticoagulant, is used.
4. The proper field specimen preparation is used.
5. The specimen is not contaminated.
6. The stability of the analyte is assured through proper shipment of the specimen to the laboratory.
7. The specimen is properly labeled.

Samples of plant and animal tissue collected under natural conditions are always contaminated with the medium in which they live. Particles of soil adhere to the roots of land plants, and sections of animal tissue are filled with blood and/or other fluids. While it may be possible to wash plant and animal tissues free of contamination from their natural media, such procedures are also likely to leach trace elements from the tissues. Red blood cells cannot be washed with water; they undergo hemolysis. Such tissues can be successfully washed with isotonic solutions made from high purity salts and high purity water. When this is done, however, the samples become contaminated with the salts used to maintain osmotic pressure.

Tissue samples are vulnerable to contamination during collection. Talc from the pathologist's gloves is frequently a source of trace element contamination. Wilkerson et al.[80] have shown that chromium, cobalt, and iron can be leached from talc by body fluids. Particles of talc falling into the sample, however, present a much more serious source of contamination. To eliminate the problem, talc-free gloves are recommended.

The implements used to collect the sample are also potential sources of contamination. Versieck and Speecke[81] have shown that during collection, blood is contaminated with chromium, cobalt, copper, iron, manganese, nickel, and zinc from disposable venipuncture needles. The same author similarly demonstrated that liver biopsy samples were contaminated with antimony, chromium, cobalt, copper, gold, manganese, nickel, scandium, silver, and zinc from the surgical blades used to excise the biopsy material. The extent of manganese contamination of the blood serum was at a level equal to its normal manganese content. Stainless steel and chromium-plated instruments in particular should be avoided. Versieck and his coworkers[49,64] have made extensive studies of the contamination of biological tissues and fluids with trace elements from radioactivated surgical instruments. They found that the chromium and nickel contents of liver tissues were doubled by contact with a scalpel blade. Some of the special precautions against chromium contamination from stainless steel have been described by Katz and Salem.[82] Plastic instruments should be used in handling the specimens to eliminate or at least minimize contamination by trace elements. Knives can be fabricated from high purity quartz or from a metal other than those for which the analyses will be conducted. High purity titanium is useful for this purpose. Katz[66] has described procedures for the collection and preservation of biological tissues and fluids for trace element analysis.

The International Atomic Energy Agency[83] has recommended the following procedure for the collection of biological specimens:

Selection of Tissues

In the WHO/IAEA [World Health Organization/International Atomic Energy Agency] study the samples selected for analysis include heart, liver, kidney cortex, and hair. Kidney medulla was also included in the original protocols but was later deleted because the large natural variability in trace element concentrations in this tissue made the interpretation of the analytical results difficult. The organs are chosen as far as possible to exclude gross localized lesions, as well as other evident inhomogeneities and pathological portions.

Heart

The sample is collected from the anterior wall of the left ventricle and must include essentially the greater part of the wall from outside to inside but exclude the endocardium, epicardium, and any subcardial fat. This site was chosen because the sample can be taken before opening the heart and thus avoids excessive contamination with blood. The interventricular system is to be avoided because of the presence of conducting tissue.

Liver:

The sample is taken from the superior anterior surface of the right lobe after removal of the capsule and of several subjacent millimeters of tissue which contain excess fibrous material.

Kidney:

A sample of kidney cortex is taken after removal of the capsule; care must be taken to avoid the cortico-medullary junction. The sample must come from the lower pole of the left kidney. A 18 mm color film is circulated among all WHO collaborating pathologists to show the standardized procedure to collect the above tissue.

Hair:

A lock of hair approximately the size of a matchstick is clipped from the occipital region, preferably using a plastic scissors.

Toenails:

Clippings as wide as possible, and comprising at least 20 mg of material are obtained from the right and left big toes. Toenails instead of fingernails were chosen because, at least in populations normally wearing shoes, they are not significantly contaminated by metals other than trace elements, whereas fingernails are exposed to occupational and other kinds of everyday contamination.

The Recommended Procedure:

Since the elements of interest are present in the tissues at concentrations of a few micrograms per gram and even lower, great care is needed to avoid metal contamination. Handling of the samples should therefore be kept to a minimum and metal-free plastic gloves should be worn. No chemical fixatives should be used and the samples should not be pierced through with a metal instrument, nor should they be rinsed with tap water or any other medium. The instruments used for handling the samples prior to analysis, or for cutting or breaking them into small pieces, may be a potent source of contamination unless the proper precautions are taken. Therefore, an autopsy collection kit comprising plastic forceps, pre-cleaned plastic vials, specially prepared titanium knives, and a stone for sharpening them is prepared at IAEA and supplied to the collecting pathologists. Silica or plastic knives have also been tested and found to introduce no significant contamination at least for certain elements, and therefore their use is acceptable. If it is necessary to use steel instruments those should preferably be of carbon steel rather than stainless steel. Stainless steel instruments, for example, forceps, scalpels, and scissors, could be used, but definitely not where chromium is to be analyzed. The danger of contaminating samples with chromium from stainless steel instruments and of course even more from chromium-plated instruments is well documented. As compared with past experience with glass knives, quartz knives, or polyethylene knives such as those used in picnic sets and stiffened by cooling with liquid nitrogen, the titanium knives recently produced at IAEA are well acceptable to pathologists, have a low degree of metal contamination, can be sharpened, and are easy to produce.

Although blood is not one of the tissues of primary interest in the present program, some consideration has to be given to the problem of obtaining suitably uncontaminated samples. Blood collecting needles made of nickel or platinum on Teflon mounts are commercially available and are considered suitable for this purpose.

Koirtyohann and Hopps[65] reviewed criteria for the selection, collection, preservation, and storage of human tissues. Littell[84] has developed a plan to use standardized sampling and analytical procedures for monitoring toxic residues in tissues of fish, and Falandysz[85] has reported on the collection and analysis of wild game to determine the presence of toxic and nontoxic trace elements.

2.4.10 Sampling Foods and Food Products

As is the case with the other sample matrices, the samples of food taken to the laboratory must be representative. Individual items of fresh fruits and vegetables or fresh meat, poultry, and fish can be collected from several points in the processing system and composited. Composite samples of solid, semisolid, and liquid food products can be obtained similarly. Individual items of canned or frozen foods can be collected from several points or sources and composited. Samples of prepared foods can be obtained as composites of individual items selected from several points in the production or distribution system. If the sample preparation is delayed, the composite should be refrigerated during storage.

Inedible or nonedible portions of the food (bones, skins, scales, rinds, husks, shells, etc,) should be removed prior to homogenization. In selecting the apparatus for handling, cutting, and homogenizing the sample, potentials for contamination and loss must be considered. Borosilicate glass vessels may be preferable to stainless steel vessels for some applications. Intermittent cooling of the vessel during homogenization of the sample may retard thermal decomposition of the sample and/or loss of volatiles.

2.5 SAMPLE PRESERVATION AND HOLDING TIMES

It is necessary to ensure that the sample undergo no changes in composition during the interval between collection and analysis. In some instances, partial loss of analyte can be prevented by the addition of preservatives. In other cases, the samples are frozen or cooled to retard such losses. Typical of such losses are the absorption of metal ions from water samples onto the walls of glass containers and the spontaneous precipitation of calcium phosphate with coprecipitation of other metals from urine samples upon standing. Acidification, cooling, and prompt analysis generally minimize such losses.

2.5.1 Preservation of Air Samples

Airborne particulates filtered from air samples are usually quite stable with respect to their trace element contents. However, Thompson[86] has indicated that fly ash can absorb mercury from contaminated laboratory air. Ehman et al.[87] reported losses by reduction to the trivalent form of approximately 40% from 5 μg hexavalent chromium spikes on fiber glass filters during the first month of postsampling storage, and Arar et al.[88] found that the reduction of hexavalent chromium in sludge incinerator particulates on quartz fiber filters followed first-order kinetics with a half-time of 21 days during the postsampling storage period. While no chromium was lost from the filters, change from the carcinogenic hexavalent form to noncarcinogenic trivalent chromium impacts risk assessments based on speciation. Air samples stored in impingers are generally stable. Nonetheless, analysis should not be delayed for prolonged periods of time.

2.5.2 Preservation of Water Samples

Both the NJ DEP[28] and the US EPA[89] require that water and wastewater samples for trace metal analysis be preserved by the addition of concentrated nitric acid immediately following their collection until the pH of the resulting solutions are reduced to at least 2. Samples so preserved may be held for a maximum of 180 days prior to the determination of aluminum, antimony, arsenic, barium, beryllium, cadmium, total chromium, copper, iron, lead, manganese, nickel, selenium, silver, tin, calcium, magnesium, potassium, and sodium. The determination of hexavalent chromium must be initiated with 24 hours of collection, and analysis for mercury may not be delayed more than 28 days when the preserved samples are stored in glass or plastic containers.

Subramanian et al.[90] have studied the loss upon storage of 11 trace metals from natural and synthetic water samples. Pyrex glass, Nalgene polyethylene, and Teflon containers were used. Trace metal losses from solutions of varying pH were evaluated with respect to storage time. Acidification to below pH 1.5 with nitric acid and storage in Nalgene containers appeared to be the most efficient in preventing losses of trace metals from natural water.

Truitt and Weber[91] found that significant amounts of copper were lost when water at pH 7 was passed through cellulose acetate filters on glass supports. Losses were minimal when acid-washed polycarbonate filters with polycarbonate supports were used. Hoyle and Atkinson[92] have reported that diammonium ethylenediamine tetraacetic acid (EDTA) prevented the absorption of lead and cadmium from aqueous solutions of pH 10 on to glass or plastic surfaces for as long as 40 days. Das et al.[93] have presented the quantitative aspects of metal ion absorption from aqueous solutions onto container surfaces. There is little doubt that metal ions can be lost from solution by absorption on to the walls of the container; Preservation is necessary. Eagle-Picher (Miami, OH), has made available a "pocket slide rule" correlating analytes, types of container for liquid and solid samples, preservations for liquid samples, and the corresponding EPA-200 and SW-846 7000 series methods for quantification.

2.5.3 Preservation of Solid Samples

Specifications for the addition of preservatives to sludge, sediment, soil, and solid waste samples are absent from regulatory compliance monitoring protocols. Preservation of solid samples is usually accomplished by cooling to 4°C. Initiation of the sample preparation should not be delayed. Holding times for prepared samples can be as long as 180 days for most metals, but delay in quantification should be avoided.

2.5.4 Preservation of Biological Fluids and Tissues

NIOSH[101] has specified that 5 mL of concentrated nitric acid be added to urine samples in polyethylene bottles prior to their refrigerated transport to the laboratory for the determination by inductively coupled plasma–atomic emission spectrometry of aluminum, barium, cadmium, chromium, copper, iron, lead, manganese, molybdenum, nickel, platinum, silver, strontium, tin, titanium, and zinc.

The addition of preservatives such as formaldehyde and alcohol to samples of biological tissues should be avoided. If the anlysis will be delayed, the sample should be lyophilized and frozen for extended storage. Preservation during short-term storage may be accomplished by cooling to 4°C.

2.6 SAMPLE COLLECTION AND PRESERVATION

Without validated and documented sample collection and preservation protocols, measurements with sophisticated laboratory equipment cannot be related to the assessment of environmental quality.

3

SAMPLE PREPARATION

Samples for elemental analysis by atomic spectrometry are usually introduced into the spectrometer as liquids. Hence, it is often necessary to dissolve airborne particulates, solid wastes, and plant and animal tissues prior to measuring the trace element concentrations. Because the atomic spectrometric techniques are comparative, the samples and standards frequently require matrix matching. Consequently, both liquid and solid samples are often treated by chemical means to better define the matrix. While atomic spectrometric techniques are sensitive and selective, additional chemical processing sometimes becomes necessary to concentrate and/or isolate the analyte. In the course of these procedures as well as those used to dissolve the sample or change its matrix, special caution must be exercised to prevent contamination. Equal care must be taken to prevent the loss of analyte.

3.1 EXTRACTION AND DECOMPOSITION FOR SOLUBILIZATION

Extraction involves removing the analyte from its matrix without destroying the matrix. While determinations made on extracts may yield low results for total analyte contents, such results are often valuable indicators of bioavailability and important for assessing environmental consequences of exposures to metal ions.

Decomposition involves removal of the organic matrix by converting it to compounds that are easily volatilized. The residue is then treated with solvent to complete solution. Among the common decomposition methods are combustion with oxygen in the presence or absence of fluxes and digestion with acids. Methods of the first type are vulnerable to volatilization losses as well as to contamination when fluxes are used, and the latter are subject to contamination with impurities in the acids. Microwave technology has enhanced both approaches.

Quicker, cleaner, safer sample preparation procedures resulted from combining the ability of microwaves to quickly heat materials with the superior dissolving properties of solvents in closed vessels under pressure. Kingston and Jassie[94] developed closed-vessel microwave technologies for decomposing in minutes samples that required hours of digestion in open vessels on hot plates. The initial cooperation between the US National Bureau of Standards [NBS—now National Institute of Science and Technology (NIST)] and the CEM Corporation was responsible for these sweeping improvements. In addition to CEM Corporation (Matthews, NC), some other vendors of microwave-assisted sample preparation systems are OI Analytical (College Station, TX) Milestone (Sorisole, Italy), Questron Corporation (Princeton, NJ), and Spex Industries, Inc. (Edison, NJ). Each markets sample preparation systems in which microwave energy is converted to thermal energy in microwave-absorbing reagents containing sample inside pressurized, microwave-transparent containers. Digestion is accelerated because pressurization allows attainment of higher temperatures (200°C in closed vessels vs. 120°C for conventional beakers-on-the-hot-plate digestions with nitric acid). Pressurization also retards evaporation of reagents and prevents volatilization losses of analyte. The former benefit has a special significance in that when less reagent is needed for the digestion, contamination of the sample with impurities in the reagent is reduced.

In addition to the microwave-assisted digestion systems, a temperature-controlled block Kjeldahl digestion system[95] (Tecator AB, Höganäs, Sweden) and an ultraviolet photolysis digestion system[96] (Metrohm Ltd., Herisau, Switzerland) have been successfully applied to the preparation of samples for atomic spectrometry.

3.1.1 TCLP and Other Extraction Procedures

The toxicity characteristic leaching procedure (TCLP), US EPA Method 1311, determines whether or not a waste is hazardous on the basis of toxicity by subjecting a sample to acid leaching conditions similar to those the waste might experience after landfill disposal. The concentrations of 40 toxicants in the leachate can then be determined. The waste is classified as hazardous on the basis of toxicity when the concentration of any toxicant in the leachate exceeds the regulatory levels. The federal regulatory levels for the eight elemental toxicants determined by atomic spectrometry are as follows: 5 mg/L arsenic, 100 mg/L barium, 1 mg/L cadmium, 5 mg/L chromium, 5 mg/L lead, 0.2 mg/L mercury, 1 mg/L selenium, and 5 mg/L silver.

For wastes with less than 0.5% solids, the liquid phase obtained by filtration through a 0.6–0.8 μm glass fiber filter is the TCLP extract.

For wastes with 0.5% solids or more, the liquid, if any, is separated by filtration through a 0.6–0.8 μm glass fiber filter and retained for the determination of the elemental toxicants just listed. The TCLP extraction of the solid phase is conducted as follows:

- Particle size reduction is required, unless the solid has a surface area of at least 3.1 cm^2/g or is smaller than 1 cm at its narrowest dimension.
- A 100 g specimen of the solid phase is weighed into the extraction vessel.

- The appropriate extraction fluid* is added to the extraction vessel containing the specimen of solid phase in an amount equivalent to 20 times the sample mass on a wet weight basis.
- The extraction vessel is closed, secured in the device for its rotation, and rotated at 30 ± 2 rpm for 18 ± 2 hours.
- The contents of the extraction vessel are separated by filtration through a 0.6–0.8 μm glass fiber filter, and the liquid phase is adjusted to pH < 2 by the addition of nitric acid in preparation for the determination of the eight elemental toxicants listed.
- If compatible, the liquid phase from the initial filtration of the waste and the TCLP extract are combined prior to the determination of the elemental toxicants. If incompatible, the liquids are analyzed separately, and the results are mathematically combined to yield a volume–mass average concentration for each of the elemental toxicants.

Smith[97] has described the toxicity characteristic leaching procedure with the flowchart reproduced in Figure 3.1.

On June 20, 1990, the toxicity characteristic leaching procedure (TCPL, Method 1311) replaced the extraction procedure toxicity (EPTOX, Method 1310) as the official method for classifying wastes as hazardous or nonhazardous on the basis of toxicity.

A variety of vendors provide extraction and filtration apparatus for the TCLP. Among them are Analytical Testing and Consulting Services, Inc. (Warrington, PA), Associated Design and Manufacturing Company (Alexandria, VA), Environmental Express (Mount Pleasant, SC), Environmental Machine and Design, Inc. (Lynchburg, VA), Fisher Scientific (Pittsburgh, PA), IRA Machine Shop and Laboratory (Santurce, Puerto Rico), Lars Lande Manufacturing (Whitmore Lake, MI), Microfiltration Systems (Dublin, CA), Millipore Corporation (Bedford, MA), Nucleopore Corporation (Pleasanton, CA), and Whatman Laboratory Products, Inc. (Clifton, NJ).

Practice B from ASTM[98] Designation D 3974–81 is capable of extracting alu-

*Determination of appropriate extraction fluid [according to US EPA Method 1311] (1) Weigh 5 g of solid phase pulverized to 1 mm or less into a 500 mL flask. (2) Add 96.5 mL of water, and stir the content of the flask for 5 minutes. (3) Measure and record the pH of the liquid phase. (a) If the pH of the liquid phase is less than 5.00, the acetic acid/sodium acetate buffer of pH 4.93 ± 0.05 is used as the extraction fluid. This buffer is prepared by treating 5.7 mL of glacial acetic acid in 500 mL of water with 64.3 mL of 1.0 N sodium hydroxide and diluting the resulting solution to 1 liter with water. (b) If the pH of the liquid phase is greater than 5.00, add 3.5 mL of 1.0 N hydrochloric acid, heat to 50°C for 10 minutes, cool, and redetermine the pH. If the pH is now less than 5.00, the acetic acid/sodium acetate buffer of pH 4.93 ± 0.05 is used as the extraction fluid. If the pH remains greater than 5.00, the acetic acid solution of pH 2.88 ± 0.05 is used as the extraction fluid. This acetic acid solution is prepared by dissolving 5.7 mL of glacial acetic acid in water and diluting the resulting solution to 1 liter with water. Extraction fluids are available from several vendors (e.g., Miller Analytical Services, Bristol, PA).

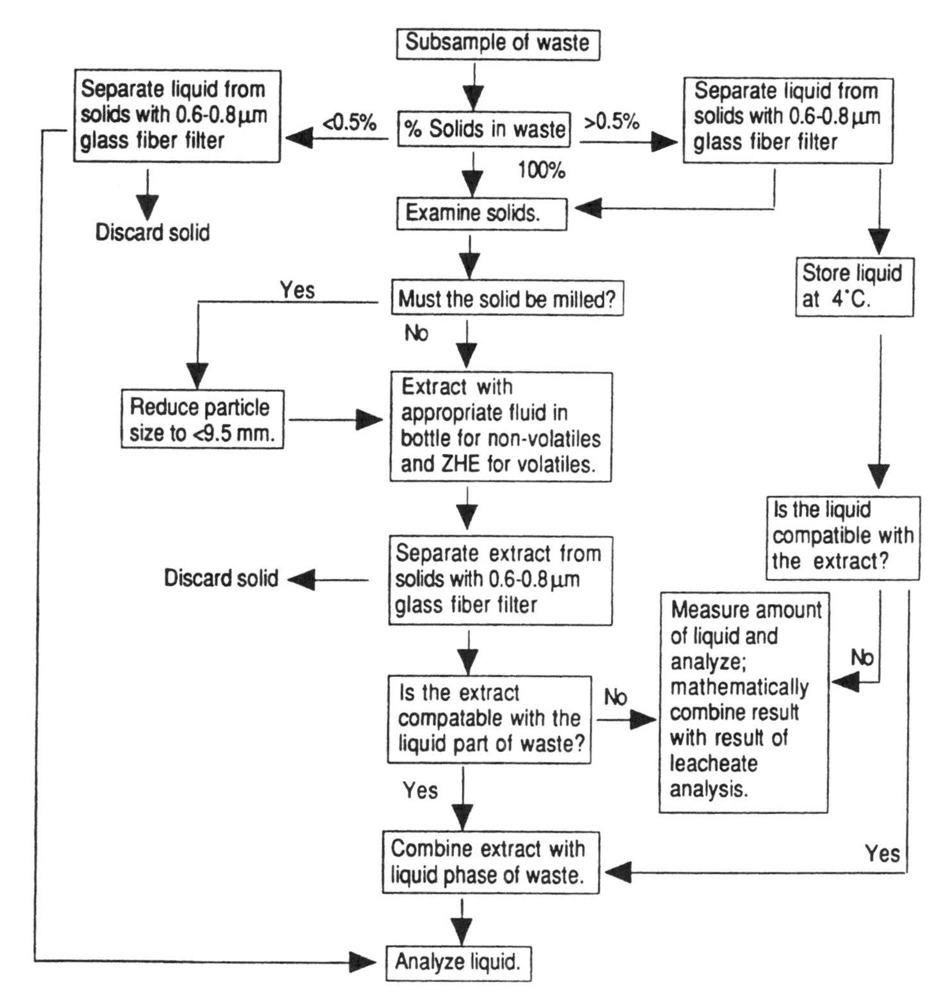

Figure 3.1 Flowchart for the toxicity characteristic leaching procedure (ZHE, Zero Headspace Extraction). *Source:* Ref. 97 with permission.

minum, cadmium, chromium, cobalt, copper, iron, lead, manganese, nickel, and zinc from soils, sediments, and waterborne particulates. The practice is as follows:

1. Weigh 1.0000 g of dried sediment and place in a 125 mL polypropylene wide-mouth bottle. For low-level trace elements use up to 10 g of sediment sample. Include an empty bottle as a reagent/glassware blank with each set of samples.
2. Add 95 mL of water and 5 mL of HCl (sp gr 1.19) to the sample and to the blank bottle and cap tightly. In the case of a foaming reaction, which indicates the presence of carbonates, add the acidic solution slowly.

3. Shake at room temperature in a mechanical shaker 16 h (overnight).
4. Filter the solution by suction filtration or filter paper. Quantitatively transfer the solution to a 100 mL volumetric flask and dilute to volume.

3.1.2 Decomposition Procedures

Decomposition involves removal of the organic matrix by converting it to compounds that are easily volatilized. The residue is then treated with solvent to complete solution. Among the common decomposition methods are combustion with oxygen in the presence or absence of fluxes and digestion with acids. The former are vulnerable to volatilization losses as well as to contamination when fluxes are used, and the latter are subject to contamination with impurities in the acids. Microwave technology has enhanced both approaches.

3.1.2.1 Fusion Procedures Practice A from ASTM[99] Designation 4698-87 is effective for the total digestion of sediments prior to the determination of aluminum, calcium, iron, magnesium, manganese, potassium, sodium, silicon, and titanium. This practice makes use of a 1:2 (m/m) mixture of $LiBO_2$ and $Li_2B_4O_7$ flux, a 4 g/L CsCl solution, and a 50 g/L H_3BO_3 solution. The practice is as follows:

10.5. Transfer approximately 1.2 g of flux mixture to a waxed or plastic-coated weighing paper [6 in. by 6 in. (152.4 mm by 152.4 mm)]. Weigh and transfer 0.2000 g of finely ground sample to the flux mixture and mix by rolling successive corners of the paper about 30 times. Carefully transfer the combined sample/flux to a graphite crucible, and tamp down by gently tapping the crucible on a tabletop.

10.6. Weigh appropriate sediment or rock standards and treat as in 10.5.

10.7. Carry several blanks through the procedure by using only flux and treat as in 10.5.

10.8. Fuse the mixtures in a muffle furnace, preheated to 1000°C for 30 min.

NOTE 2: When the crucibles, samples, and crucible racks are placed in the muffle furnace, the temperature may drop as much as 200°C. Time is still measured from the time of insertion in the furnace.

10.9. Remove the crucibles from the furnace and allow to cool; dislodge the beads by gentle tapping or with a spatula.

NOTE 3: The beads may be dissolved immediately after cooling, or can be stored in plastic vials for dissolution at a later time.

10.10. Place the bead in an acid-washed 250 mL plastic bottle and add a 0.25 to 1 in. (19.05 to 25.4 mm) magnetic stirring bar. Add approximately 50 mL boiling water using a plastic graduate, place the bottle on a magnetic stirrer, and mix. Add 5 mL of HNO_3 (1+1) to each bottle and stir rapidly for about 60 min. Cap the bottle tightly to prevent both contamination and possible spattering.

10.11. Immediately after 60 min, remove the bottles from the stirrers, and add about 100 mL of water to prevent polymerization of silica.

NOTE 4: The solution may contain small amounts of graphite from the crucibles which can be ignored. However, if the solution is cloudy, this indicates a very high concentration of silica in the original sample and that it has polymerized. Such a solution must be discarded, and a new fusion performed using a smaller quantity of sample.

10.12. Pour each solution into a 200 mL volumetric flask, using a funnel, in order to retain the stirring bar. Rinse the bottle and cap, and bring to the mark with water. Pour the solution back into the plastic bottle for storage.

10.13. Add 10 mL of CsCl solution and 20 mL of H_3BO_3 solution to each bottle.

NOTE 5: The CsCl acts as an ionization suppressant and the H_3BO_3 stabilizes the silica; these are used when quantitation is by flame atomic absorption spectrometry.

The US DOE methods[100] prescribe lithium metaborate and sodium peroxide fusions to prepare soil samples for the determination by inductively coupled plasma–mass spectrometry of ^{230}Th and ^{234}U and of ^{99}Tc, respectively.

3.1.2.2 Digestion Procedures With and Without Microwaves Sample preparation frequently begins with acid decomposition. Filters containing airborne particulates, sludges, sediments, soils, liquids containing suspended solids, and plant and animal tissues are dissolved by decomposing the sample with acid under a wide variety of experimental conditions.

Air Samples Practically all the procedures for preparing airborne particulate material for atomic spectrometry begin with the acid digestion of the filter that served to collect the sample. Cellulose ester filters are usually dissolved in nitric acid. Sometimes, hydrochloric acid, sulfuric acid, and/or perchloric acid is added to aid in the destruction of organic material. Hydrogen peroxide has also been used for this purpose. Hydrofluoric acid is required to dissolve glass fiber filters.

The ASTM[46] has proposed a procedure for decomposing cellulose ester filter membranes and dissolving particulate samples collected from workplace atmospheres prior to determining aluminum, barium, bismuth, calcium, cadmium, chromium, cobalt, copper, indium, iron, lead, lithium, magnesium, manganese, nickel, potassium, rubidium, silver, sodium, strontium, titanium, vanadium, and zinc by flame atomic absorption spectrometry. The procedure is as follows.

Blank filters and filters containing samples in a 1:10 ratio are transferred to 125 mL beakers and treated with sufficient concentrated nitric acid to cover them. The beakers are covered with watch glasses, and their contents are heated (140°C) on a hot plate in a fume hood until the samples are dissolved and pale yellow solutions are obtained. This is usually accomplished in approximately 30 minutes, but longer heating and additional nitric acid may be required for samples with high organic contents. Care must be taken to prevent evaporation to dryness. Some compounds of aluminum, chromium, and cobalt may require alternate procedures that make use of hydrochloric–nitric–sulfuric acid mixtures or hydrochloric–nitric–perchloric mixtures for dissolution. All the organic material must be destroyed before perchloric acid is added, and special care must be taken to prevent evaporation to dryness. The solutions are quantitatively transferred to 10 mL volumetric flasks, treated with Cs ionization buffers and La-releasing agents as appropriate, and brought to volume with high-purity water.

The *NIOSH Manual of Analytical Methods*[101] recommends 4:1 (v/v) nitric acid–perchloric acid for decomposing cellulose ester filter membranes and dissolving metals present in samples of airborne particulates prior to the quantification of aluminum, arsenic, beryllium, cadmium, chromium, cobalt, copper, iron, lead, lithium, magnesium, manganese, molybdenum, nickel, phosphorus, platinum, selenium, silver, sodium, tellurium, thallium, titanium, vanadium, yttrium, zinc, and zirconium by inductively coupled plasma–atomic emission spectrometry. The sample preparation procedure is as follows.

Five milliliters of 4:1 (v/v) nitric acid–perchloric acid is added to beakers containing filters on which samples have been collected, to beakers containing unused filters, and to empty beakers serving as reagent blanks. The beakers are covered with watch glasses, and their contents are heated on a hot plate (120°C) until ~0.5 mL of liquid remains. The beakers are cooled to room temperature, and the contents of each are treated with 2 mL of 4:1 nitric acid–perchloric acid and returned to the hot plate. Heating is resumed until ~0.5 mL of liquid remains. Additional 2 mL increments of acid mixture followed by heating and cooling cycles may be needed to obtain clear solutions. Recovery of lead from paint matrices and some species of aluminum, beryllium, chromium, cobalt, manganese, molybdenum, vanadium, and zirconium may require alternative solubilization techniques. The undersides of the watch glasses are rinsed with distilled water, and the rinsings are collected in the corresponding beakers. The beakers are heated until the contents are nearly evaporated to dryness (i.e., ~0.5 mL of liquid remains). The residues are dissolved with a few milliliters of 1% perchloric acid–4% nitric acid mixture, and the solutions are quantitatively transferred to 10 mL volumetric flasks. The contents of the volumetric flasks are brought to volume with 1% perchloric acid–4% nitric acid mixture.

The microwave digestion procedure for lead in paint chips described under "Miscellaneous Samples" at the end of Section 3.1.2.2 is an acceptable alternative to the beaker–hot plate procedure just described. Microwave digestion provides rapid, complete solubilization of airborne particulates collected on cellulose ester filter membranes prior to quantification by flame or furnace atomic absorption spectrometry or by inductively coupled plasma–atomic emission spectrometry.

Pundyn and Smith[102] have compared microwave-assisted digestions to beakers-on-the-hot-plate methods for the preparation of samples from workplace atmospheres containing dust, fly ash, and paint mist prior to the determination of airborne metal concentrations by inductively coupled plasma–atomic emission spectrometry. Some of their results for the NIST standard reference material coal fly ash (SRM 1633a) are reproduced in Table 3.1. The beakers-on-the-hot-plate digestions were made by heating the samples to 110°C in PTFE beakers with 3 mL of HF for 2 hours followed by heating with 10 mL of aqua regia for 3 hours. The microwave-assisted digestions were made with 5 mL of HF and 10 mL of aqua regia at power settings of 200 W for 5 minutes followed by 5 minutes at 400 W power settings. Certi-

TABLE 3.1 Comparison of Hot Plate and Microwave-Assisted Digestions for Elemental Analysis of Airborne Particulates, NIST SRM 1633a

Element	Hot Plate Digestion (5 h)	Microwave Digestion (10 min)	INAA[a]	Certified Range
Aluminum, %	13.0 ± 0.48	14.2 ± 0.50	15.0 ± 0.60	14.0–14.8
Chromium, ppm	180 ± 16	182 ± 15	210 ± 10	180–200
Cobalt, ppm	52 ± 7	56 ± 8	44 ± 2	
Iron, %	8.95 ± 0.32	9.44 ± 0.30	9.55 ± 0.21	9.50–9.60
Manganese, ppm	180 ± 23	205 ± 18	200 ± 20	170–270
Nickel, ppm	154 ± 11	158 ± 10	70 ± 10	120–190
Titanium, %	0.76 ± 0.10	0.79 ± 0.09	0.74 ± 0.09	0.78–0.90
Vanadium, ppm	280 ± 13	290 ± 12		280–360
Zinc, ppm	230 ± 22	225 ± 26	300 ± 40	220–250

[a]Instrumental neutron-activation analysis.
Source: Ref. 102.

fied values and results obtained by instrumental neutron activation analysis are also included in Table 3.1. This example clearly demonstrates the time-saving superiority of microwave-assisted digestion.

Water and Wastewater Samples Samples of water and wastewater are frequently subjected to acid digestion to dissolve suspended material, to destroy dissolved organic material, and to define the anionic medium for matrix matching.

US EPA Method 200.2[103] contains a procedure to prepare samples of drinking water, groundwater, surface water, and wastewater for the determination of "total recoverable elements" by atomic spectrometry. The procedure is as follows.

> For the determination of total recoverable elements in water or wastewater, take a 100 mL (±1 mL) aliquot from a well mixed, acid preserved sample containing not more than 0.25% (w/v) total solids and transfer to a 250 mL griffin beaker. (If total solids are greater than 0.25% reduce the size of the aliquot by a proportionate amount.) Add 2 mL of (1+1) nitric acid and 1 mL of (1+1) hydrochloric acid. Heat on a hot plate at 85°C until the volume has been reduced to approximately 20 mL, ensuring that the sample does not boil. (A spare beaker containing approximately 20 mL of water can be used as a gauge.)* Cover the beaker with a watch glass and reflux for 30 min. Slight boiling may occur but vigorous boiling should be avoided. Allow to cool and quantitatively transfer to either a 50 mL volumetric flask or a 50 mL class A stoppered graduated cylinder. Dilute to volume with ASTM type I water and mix. Centrifuge the sam-

**Note:* For proper heating, adjust the temperature control of the hot plate such that an uncovered beaker containing 50 mL of water located in the center of the hot plate can be maintained at a temperature of 85°C. Evaporation time for 100 mL of sample at 85°C is approximately two hours with the rate of evaporation increasing rapidly as the sample volume approaches 20 mL.

ple or allow to stand overnight to separate insoluble material. The sample is now ready for analysis by either inductively coupled plasma–atomic emission spectrometry or direct aspiration flame and stabilized temperature graphite furnace–atomic absorption spectroscopy. For analysis by inductively coupled plasma–mass spectrometry, pipette 20 mL of the prepared solution into a 50 mL volumetric flask, dilute to volume with ASTM type I water and mix. (Internal standards are added at the time of analysis.) Because the effects of various matrices on the stability of diluted samples cannot be characterized, all analyses should be performed as soon as possible after the completed preparation.

Method 200.2 is applicable to the preparation of drinking water, groundwater, surface water, and wastewater samples for the determination of aluminum, antimony, arsenic, barium, beryllium, boron, cadmium, calcium, chromium, cobalt, copper, iron, lead, lithium, magnesium, manganese, mercury, molybdenum, nickel, phosphorus, potassium, selenium, silica, silver, sodium, strontium, thallium, thorium, tin, uranium, vanadium, and zinc. Method 200.2 contains a cautionary note on the loss of barium by precipitation of barium sulfate from solutions containing sulfate. Lead should be similarly affected.

US EPA[104] Method 200.8 provides procedures for the preparation of groundwater, surface water, and drinking water samples prior to the determination of total recoverable concentrations of aluminum, antimony, arsenic, barium, beryllium, cadmium, chromium, cobalt, copper, lead, manganese, mercury, molybdenum, nickel, selenium, silver, thallium, thorium, uranium, vanadium, and zinc by inductively coupled plasma–mass spectrometry. Method 200.8 is also applicable to the determination of total recoverable concentrations of these elements in wastewaters, sludges, and soils. For the determination of total recoverable concentrations, an accurately measured or weighed aliquot of the well-mixed, homogeneous liquid or solid sample is gently refluxed with nitric and hydrochloric acid and brought to volume as described earlier. Spike recoveries from drinking water samples ranged from 94% for cobalt to 115% for aluminum. In sewage treatment plant effluents, spike recoveries ranged from 79% for barium and silver to 112% for selenium, and in hazardous waste site soils, spike recoveries ranged from 69% for antimony to 109% for vanadium.

The US EPA Contract Laboratory Program[105] describes three hot plate digestion procedures and one microwave-assisted digestion procedure to prepare water samples for the determination of "total metals" by atomic spectrometry.

The following procedure is required for the determination of aluminum, antimony, arsenic, barium, beryllium, cadmium, calcium, chromium, cobalt, cooper, iron, lead, magnesium, manganese, nickel, potassium, selenium, silver, sodium, thallium, vanadium, and zinc by electrothermal atomization–atomic absorption spectrometry:

Shake sample and transfer 100 mL of well-mixed sample to a 250 mL beaker, add 1 mL of (1+1) HNO_3 and 2 mL of 30% H_2O_2 to the sample. Cover with watch glass or similar cover and heat in a steam bath or hot plate for 2 hours at 95°C or until sample volume is reduced to between 25 and 50 mL, making certain the sample does not boil. Cool sample and filter to remove insoluble material. (NOTE: In place of filtering, the sample, after dilution and mixing, may be centrifuged or allowed to settle by gravity overnight to remove insoluble material.) Adjust sample volume to 100 mL with deionized distilled water. The sample is now ready for analysis.[105]

For the determination of the metals just listed by inductively coupled plasma–atomic emission spectrometry or by flame atomic absorption spectrometry, hot plate digestion with aqua regia is specified.

Shake sample and transfer 100 mL of well-mixed sample to a 250 mL beaker, add 1 mL of (1+1) HNO_3 and 10 mL of (1+1) HCl to the sample. Cover with watch glass or similar cover and heat in a steam bath or hot plate for 2 hours at 95°C or until sample volume is reduced to between 25 and 50 mL, making certain the sample does not boil. Cool sample and filter to remove insoluble material. (NOTE: In place of filtering, the sample, after dilution and mixing, may be centrifuged or allowed to settle by gravity overnight to remove insoluble material.) Adjust sample volume to 100 mL with deionized distilled water. The sample is now ready for analysis.[105]

When the metals listed are determined by inductively coupled plasma–mass spectrometry, the following procedure is required for preparation of the samples:

Shake sample and transfer 100 mL of well-mixed sample to a 250 mL beaker, add 1 mL of (1+1) HNO_3 and 2 mL of 30% H_2O_2 to the sample. Cover with watch glass or similar cover and heat in a steam bath or hot plate for 2 hours at 95°C (temperature should be monitored with a thermometer) or until sample volume is reduced to between 25 and 50 mL, making certain the sample does not boil. Cool sample and filter to remove insoluble material. Adjust sample volume to 100 mL with [ASTM] Type I water. The sample is now ready for analysis.[105]

This procedure must be used for water samples containing more than 30 μg/L of silver or more than 100 μg/L of antimony. Inductively coupled plasma–mass spectrometry may not be employed for the determination of calcium, magnesium, potassium, and sodium.

The microwave-assisted digestion procedure is applicable to the preparation of water samples for the determination of aluminum, antimony, arsenic, barium, beryllium, cadmium, calcium, chromium, cobalt, copper, iron, lead, magnesium, manganese, nickel, potassium, selenium, silver, sodium, thallium, vanadium, and zinc by electrothermal atomization–atomic absorption spectrometry, inductively coupled plasma–atomic emission spectrometry, or inductively coupled plasma–mass spectrometry. The US EPA's required procedure is as follows:

45 mL of sample are measured into Teflon digestion vessels using volumetric glassware. 5 mL of high purity HNO_3 are added to the digestion vessels and the weight recorded to 0.02 g.

The caps with the pressure release valves are placed on the vessels and then tightened, using constant torque, to 12-ft lbs. Place 5 sample vessels in the carousel, evenly spaced around its periphery in the microwave unit. Venting tubes connect each sample vessel with a collection vessel. Each sample vessel is attached to a clean, double-ported vessel to collect any sample expelled from the sample vessel in the event of overpressurization. Assembly of the vessels into the [carousel] may be done inside or outside the microwave. This procedure is energy balanced for five 45 mL water samples (each with 5 mL of acid) to produce constant conditions and prevent alteration of the conditions. The initial temperature of the sample should be 24 ± 1°C. Blanks must have 45 mL of deionized water and the same amount of acid to be added to the microwave as a reagent blank.

Power Programming of Nitric Acid: The 5 samples of 45 mL of water and 5 mL of nitric acid are irradiated for 10 minutes at 545 W and immediately cycled to the second program for 10 minutes at 344 W (BASED ON THE CALIBRATION OF THE MICROWAVE UNIT AS PREVIOUSLY DESCRIBED). This program brings the samples to 160° ± 4°C in 10 minutes and then causes a slow rise in temperature between 165 [and] 170°C during the second 10 minutes.

Following the 20 minute program, the samples are left to cool in the microwave unit for 5 minutes, with the exhaust fan ON. The samples and/or carousel may then be removed from the microwave unit. Before opening the vessels, let cool until they are no longer hot to the touch.

After the sample vessel has cooled, weigh the sample vessel and compare it to the initial weight as reported in the preparation log. Any vessel exhibiting a ≤ 0.5 g loss must have any excess sample from the associated collection vessel added to the original sample before proceeding with the sample preparation. Any sample exhibiting a > 0.5 g loss must be identified in the preparation log and the sample redigested.

Sample Filtration: The digested samples are shaken well to mix in any condensate within the digestion vessel before being opened. The digestates are then filtered into 50 mL glass volumetric flasks through ultra-clean filter paper and diluted to 50 mL (if necessary). The samples are now ready for analysis. The results must be corrected by a factor of 1.11 in order to report the concentration values based on an initial volume of 45 mL. Concentrations so determined shall be reported as "total."[105]

The 1.11 correction factor is multiplicative, and it is derived from 50.0 ÷ 45.0 = 1.11. It should be noted from the procedures quoted that hot plate digestions require 2 hours while microwave-assisted digestions are complete in 25 minutes. The calibration procedure for the microwave unit is as follows:

The calibration procedure is a critical step prior to the use of any microwave unit. In order that absolute power settings may be interchanged from one microwave unit to another, the actual delivered power must be determined.

Calibration of a laboratory microwave unit depends on the type of electronic system used by the manufacturer. If the unit has a precise and accurate linear relationship between the output power and the scale used in controlling the microwave unit, then the calibration can be a single-point calibration at maximum power. If the unit is not accurate or precise for some portion of the controlling scale, then a multiple point calibration is necessary. If the unit power calibration needs multiple point calibration, the point where linearity begins must be identified. For example: a calibration at 100, 99,

98, 97, 95, 90, 80, 70, 60, and 50% power settings can be applied and the data plotted. The nonlinear portion of the calibration curve can be excluded or restricted in use. Each percent is equivalent to approximately 5.5–6.5 W and becomes the smallest unit of power that can be controlled. If 20–40 W are contained from 99–100%, that portion of the microwave calibration is not controllable by 3–7 times that of the linear portion of the control scale and will prevent duplication or precise power conditions specified in that portion of the power scale.

The following equation evaluates the power available for heating a microwave cavity. This is accomplished by measuring the temperature rise of 1 kg of water exposed to electromagnetic radiation for a fixed period of time. Measurements are made on weighed replicates (5 replicates) of one kilogram samples of room temperature distilled water in thick-walled microwave transparent (Teflon) vessels. The containers must be circulated continuously through the field for at least two (2) minutes at full power. The vessel(s) are removed from the microwave and stirred. After stirring, the temperature of the water is measured and recorded for use in the formula below. One kilogram of water in one container or equally divided between two containers is acceptable.

Calibration Formula: Weighed replicates (5) of one kilogram distilled room-temperature water in a microwave transparent vessel:

Measure initial temperature of water (T_i) to within 0.1°C. The starting temperature should be between 22 and 26°C.

Irradiate 1 kilogram of water at full power, 100% (99, 98, 97, 95, 90, 80, 70, 60, or 50% power setting) for 120 seconds. The container must be circulated through the cavity at a rate of at least one revolution every 30 sec. during the irradiation.

Measure the final temperature of water (T_f), after stirring, to within 0.1°C with stirring (an electronic stirrer using a large stir bar works best) within 30 sec of the end of microwave irradiation. Take the maximum reading.

Repeat for a new sample, for a total of five replicates per microwave setting, of distilled room-temperature water on [sic] a new clean container.

Calculate microwave power according to the formula:

$$\text{Power} = \frac{K \times C_p \times M \times T}{t} \qquad (T_f - T_i) = T$$

$$\frac{K \times C_p \times M \times T}{t} = 34.87$$

$$\text{Power} = 34.87 \times T$$

where:

Power = the apparent power absorbed by the sample in watts ($W = \text{joule}\cdot\text{sec}^{-2}$);
K = the conversion factor for thermochemical calories·sec^{-1} to W (= 4.184);
C_p = the heat capacity, thermal capacity, or specific heat, ($\text{cal}\cdot\text{g}^{-1}\cdot{}^\circ\text{C}^{-1}$ = 1.0 for water);
M = mass of the sample in grams (g); $T = (T_f - T_i)$ in °C;
t = time in seconds (s).

Derive an equation for the linear portion of the calibration range and determine the equivalent value in watts for the arbitrary setting scale. Use the actual power in watts to determine the appropriate setting of the particular microwave unit being used. Each microwave unit will have its own setting that corresponds to the actual power delivered to the samples.[105]

Alternative documentation[106] states: "The microwave unit must be calibrated every six months."

Oily Waste Samples The ASTM Standard Practice for Nitric Acid Digestion for Solid Waste[107] (D-5198-92) also has been applied to oily sludges for the solubilization of aluminum, beryllium, cadmium, chromium, copper, iron, lead, manganese, mercury, nickel, phosphorus, vanadium, and zinc. In addition, this practice may be applicable to solubilization of arsenic, barium, calcium, cobalt, magnesium, and selenium; but silver, silicon, and titanium are not solubilized by this practice.

Solid Waste, Sludge, Sediment, and Soil Samples US EPA Method 200.8, described under "Water and Wastewater Samples" in Section 3.1.2.2, is also applicable to sludges, sediments, and soils.

US EPA Method 3050, Acid Digestion of Sediments, Sludges, and Soils, differs only slightly from those described earlier to prepare water and wastewater samples for the determination of metals by inductively coupled plasma–mass spectrometry. An accurately weighed (1–2 g) specimen of solid is treated with 10 mL of 1+1 HNO_3 and heated. Two additional increments of HNO_3 are added, and heating is continued to decompose the organic matrix. Further decomposition of the organic matrix is accomplished by continued heating and the addition of up to five 2 mL increments of 30% H_2O_2. After cooling, insoluble material is removed by filtration, and the filtrate is diluted to 100 mL in a volumetric flask.

Method 3050 has undergone interlaboratory study.[108–111] Method 6010 performance data for 23 elements in seven different solid waste matrices prepared by Method 3050 indicated that 55% of the overall variance was due to sample preparation, with 45% attributable to quantification by inductively coupled plasma–mass spectrometry.

An alternative to US EPA Method 3050 is US EPA Method 3051,[112] Microwave Assisted Acid Digestion of Sediments, Sludges, Soils, and Oils. Method 3051 is intended to provide a rapid acid extraction prior to atomic spectrometric quantification of aluminum, antimony, arsenic, barium, beryllium, boron, cadmium, calcium, chromium, cobalt, copper, iron, lead, magnesium, manganese, mercury, molybdenum, nickel, potassium, selenium, silver, sodium, strontium, thallium, vanadium, and zinc. The microwave equipment must be calibrated prior to use by the procedure described earlier. The sample digestion procedure is as follows:

7.3.1. Weigh the fluorocarbon [polytetrafluoroethylene] digestion vessel, valve and cap assembly prior to use.

7.3.2. Weigh a well-mixed sample to the nearest 0.001 g into the fluorocarbon vessel equipped with a single-ported cap and a pressure relief valve. For soils, sediments, and sludges use no more than 0.500 g. For oils use no more than 0.250 g.

7.3.3. Add 10 ± 0.1 mL concentrated nitric acid in a fume hood. If a vigorous reaction occurs, allow the reaction to stop before capping the vessel. Cap the vessel and torque the cap to 12 ft-lbs (16 N-m) or according to the unit manufacturer's directions. Weigh the vessel to the nearest 0.001 g. Place the vessel in the microwave carousel.

CAUTION: Toxic nitrogen oxide fumes may be evolved, therefore all work must be performed in a properly operating ventilation system. The analyst should also be aware of the potential for a vigorous reaction. If a vigorous reaction occurs, allow to cool before capping the vessel.

CAUTION: When digesting sample containing volatile or easily oxidized organic compounds, initially weigh no more than 0.10 g and observe the reaction before capping the vessel. If a vigorous reaction occurs, allow the reaction to cease before capping the vessel. If no appreciable reaction occurs, a sample weight up to 0.25 g can be used.

CAUTION: All samples known or suspected of containing more than 5–10% organic material should be predigested in a hood for at least 15 minutes.

7.3.4. Properly place the carousel in the microwave unit according to the manufacturer's recommended specifications and, if used, connect the pressure vessels to the central overflow vessel with PFA–fluorocarbon tubes. Any vessels containing 10 mL of nitric acid for analytical blank purposes are counted as sample vessels. When fewer than the recommended number of samples are to be digested, the remaining vessels should be filled with 10 mL of nitric acid to achieve the full complement of vessels. This provides an energy balance since the microwave power absorbed is proportional to the total mass in the cavity [Ref. 4]. Irradiate each group of sample vessels for 10 minutes. The temperature of each vessel should rise to 175°C in less than 5.5 minutes and remain between 170 [and] 180°C for the balance of the 10 minute irradiation period. The pressure should peak at less than 6 atm for most soil, sludge, and sediment samples [Ref. 5]. The pressure will exceed these limits in the case of high concentrations of carbonate or organic compounds. In these cases the pressure will be limited by the relief pressure of the vessel to 7.5 ± 0.7 atm (110 ± 10 psi). All vessels should be sealed according to the manufacturer's specifications.

Newer microwave units are capable of higher power (W) that permits digestion of a larger number of samples per batch. If the analyst wishes to digest more samples at a time, the analysis may use different values of power as long as they result in the same time and temperature conditions defined in 7.3.4. That is, any sequence of power that brings the samples to 175°C in 5.5 minutes and permits a slow rise to 175–180°C during the remaining 4.5 minutes [Ref. 5].

Issues of safety, structural integrity (both temperature and pressure limitations), heating loss, chemical compatibility, microwave absorption of vessel material, and energy transport will be considerations made in choosing alternate vessels. If all of the considerations are met and the appropriate power settings provided to reproduce the reaction conditions defined in 7.3.4, then these alternate vessels may be used [Ref. 1,2].

7.3.5. At the end of the microwave program, allow the vessels to cool for a minimum of 5 minutes before removing them from the microwave unit. When the vessels have cooled to room temperature, weigh and record the weight of each vessel assembly. If the weight of acid plus sample has decreased by more than 10 percent from the original weight, discard the sample. Determine the reason for the weight loss. These are typically attributed to loss of vessel seal integrity, use of a digestion time longer than 10 minutes, too large a sample, or improper heating conditions. Once the source of error has been corrected, prepare a new sample or set of samples beginning at 7.3.1.

7.3.6. Complete the preparation of the sample by carefully uncapping and venting each vessel in a fume hood. Transfer the sample to an acid-cleaned bottle. If the digested sample contains particulates which may clog nebulizers or interfere with injection of the sample into the instrument, the sample may be centrifuged, allowed to settle, or filtered.

Centrifugation: Centrifugation at 2,000–3,000 rpm for 10 minutes is usually sufficient to clear the supernatant.

Settling: Allow the sample to stand until the supernatant is clear. Allowing the sample to stand overnight will usually accomplish this. If it does not, centrifuge or filter the sample.

Filtering: The filtering apparatus must be thoroughly cleaned and prerinsed with dilute (approximately 10% v/v) nitric acid. Filter the sample through qualitative filter paper into a second acid-cleaned container.

7.3.7. Dilute the digest to a known volume assuring that the samples and standards are matrix matched. The digest is now ready for analysis for elements of interest using the appropriate SW-846 methods.

Typical recovery and bias data for the determination of some wear metals in an oil sample (SRM 1085) prepared by Method 3051 are presented in Table 3.2.

Zehr et al.[113] have compared microwave-assisted digestions with 10% (m/v) LiOH/30% H_2O_2 solution to fusion with Na_2CO_3/Na_2O_2 mixture for the preparation of tungsten carbide–tungsten oxide scrap prior to determining the vanadium content. While recoveries of vanadium spikes were recorded at 94% from the scrap samples prepared by the former and 70% from the latter, the results for the samples showed no significant differences. Nonetheless, the authors cite decreased sample

TABLE 3.2 Recovery and Bias Data for Some Ware Metals in Determined in Oil SRM 1085 Prepared by US EPA Method 3051

	Recovery (mg/kg)			
Element	Experimental	Certified	Bias (%)	P_{95}[a]
Aluminum	295 ± 12	296 ± 4	0	No
Chromium	293 ± 10	298 ± 5	−2	No
Copper	289 ± 9	295 ± 20	−2	No
Iron	311 ± 14	300 ± 4	+4	No
Magnesium	270 ± 11	297 ± 3	−9	Yes
Molybdenum	238 ± 11	292 ± 11	−18	Yes
Nickel	293 ± 9	303 ± 7	−3	No

[a]Bias significant at 95% level.

preparation time and lower risk of contamination as advantages of microwave-assisted digestions with the $LiOH/H_2O_2$ solution.

Botanical and Biological Samples US EPA Method 200.3[114] is applicable to the preparation of biological tissue samples prior to using atomic spectrometry for quantifications of aluminum, antimony, arsenic, barium, beryllium, cadmium, calcium, chromium, cobalt, copper, iron, lead, lithium, magnesium, manganese, mercury, molybdenum, nickel, phosphorus, potassium, selenium, silver, sodium, strontium, thallium, thorium, uranium, vanadium, and zinc. Like Method 3050, Method 200.3 is a beaker-on-the-hot-plate digestion with nitric acid and hydrogen peroxide. The procedure is as follows:

> 11.1. Place up to a 5 g subsample of frozen tissue into a 125 mL Erlenmeyer flask. Any sample spiking solutions should be added at this time and allowed to be in contact with the sample prior to the addition of acid.
>
> 11.2. Add 10 mL of concentrated nitric acid and warm on a hot plate until the tissue is solubilized. Gentle swirling of the sample or use of an oscillating hot plate will aid in this process.
>
> 11.3. Increase the temperature to near boiling until the solution begins to turn brown. Cool sample, add an additional 5 mL of concentrated nitric acid and return to the hot plate until the solution once again begins to turn brown.
>
> 11.4. Cool sample, add an additional 2 mL of concentrated nitric acid, return to the hot plate and reduce the volume to 5–10 mL. Cool sample, add 2 mL of 30% hydrogen peroxide, return sample to the hot plate and reduce the volume to 5–10 mL.
>
> 11.5. Repeat section 11.4 until the solution is clear or until a total of 10 mL of peroxide has been added. NOTE: A laboratory reagent blank is especially critical in this procedure because the procedure concentrates any reagent contaminants.
>
> 11.6. Cool the sample, add 2 mL of concentrated hydrochloric acid, return to the hot plate and reduce the volume to 5 mL.
>
> 11.7. Allow the sample to cool and quantitatively transfer to a 100 mL volumetric flask. Dilute to volume with ASTM type I water, mix, and allow any insoluble material to separate. The sample is now ready for either ICP-AES or STGFAA [stabilized temperature, graphite furnace–flame atomic absorption spectrometry]. For analysis by ICP-MS an additional dilution (1 + 4) is required.

An alkaline digestion with tetramethylammonium hydroxide is described in US EPA Method 200.11[115] for the preparation of fish tissue prior to the quantification of the micro- and macroelement contents by ICP-AES. Elements determined in fish tissues prepared by Method 200.11 include aluminum, antimony, arsenic, beryllium, cadmium, calcium, chromium, copper, iron, lead, magnesium, nickel, phosphorus, potassium, selenium, sodium, thallium, and zinc.

Specimens of fish tissue (1–2 g) cut from frozen fillets are accurately weighed into 30 mL polysulfone centrifuge tubes. A volume in milliliters of 25% aqueous tetramethylammonium hydroxide equal numerically to the mass in grams of the specimen is added to each tube. The tubes are securely closed and heated in a 65 ±5°C oven for an hour. After this time the tubes are removed from the oven, vortexed briefly, and returned to the oven for an additional hour. After the second heating, the tubes are again vortexed and allowed to stand overnight.

On the following morning, concentrated nitric acid is added to the contents of each tube. The volume of acid added to each tube is determined from the original mass of the specimen as follows.

Mass of Specimen (g)	Volume of Acid (mL)
0.80–104	0.4
1.05–1.24	0.5
1.25–1.44	0.6
1.45–1.64	0.7
1.65–1.84	0.8
1.85–2.04	0.9
2.05–2.24	1.0

After addition of the acid, the tubes are capped tightly and vortexed briefly, then heated in a 100°C oven for an hour. Proteinaceous material is precipitated.

The tubes containing the precipitate are removed from the oven and allowed to cool. Distilled water is added to the contents of each in an amount such that the final volume will be 10 times the original specimen mass. After dilution, the contents are vortexed briefly and then centrifuged at 2000 rpm for 10 minutes. With special care not to disturb the precipitates, the supernatants are aspirated into the ICP atomic emission spectrometer.

Results for several major and minor elements obtained from specimens of oyster tissue (SRM 1566) prepared by alkaline digestion showed recoveries ranging from 66% for iron to 124% for nickel. Murthy et al.[116] found earlier than 10% aqueous tetramethylammonium hydroxide was as efficient as nitric acid for the digestion of liver and kidney samples. Julshamn and Anderson[117] have recommended Lumaton, a proprietary tetraalkylammonium hydroxide in toluene, for the digestion of muscle biopsy samples. The concentrations of copper, cadmium, and manganese determined in specimens of the NBS bovine liver (SRM 1577) digested with this reagent agreed with the reference values and values obtained from specimens of this material digested with nitric acid.

NIOSH Method 8005[118] makes use of a 3:1:1 v/v/v mixture of nitric, perchloric, and sulfuric acids for the beaker-on-the-hot-plate digestion of blood and other biological tissues. The sample preparation is as follows:

> Allow the sample to equilibrate to room temperature.
>
> Transfer accurately weighed portions of 10 mL blood, 0.25 g "dry" tissue, or 1.0 g "wet" tissue to a beaker.
>
> Add 10.0 mL digestion acid to each blood sample or 5.0 mL digestion acid to each tissue sample. Heat at 110°C for 2 h.
>
> NOTE: Start reagent blanks, in triplicate, at this step.

Increase hotplate temperature to 250°C and heat until ca. 1 mL (for blood) or ca. 0.5 mL (for tissues) remains (2 to 3 h).

Allow beakers to cool.

Choose one of the following:

(a) External standard method: Transfer contents of beaker to a volumetric flask (10 mL for blood, 5 mL for tissue). Dilute to the mark with deionized water.

(b) Internal standard method: Add, via pipet, 10.0 mL (for blood) or 5.0 mL (for tissue) yttrium standard to beaker.

Recoveries for 19 metals from blood ranged from 81% for cobalt to 131% for vanadium. Copper and zinc recoveries were 101 and 103%, respectively.

Miscellaneous Samples Beaker-on-the-hot-plate digestion has been compared to microwave-assisted digestion of paint chips for the determination of lead by atomic absorption spectrometry and inductively coupled plasma–atomic emission spectrometry.[119] The former employed the nitric acid–hydrogen peroxide mixture of Method 3050 described earlier, while the latter made use of a solution 0.9 M in nitric acid and 2.1 M in hydrochloric acid.

ASTM Designation D 5513-94 describes a microwave digestion of industrial furnace feedstreams derived from coal, coke, cement raw feeds, and fuels composed of paint-waste-related materials using sequential additions of nitric, hydrofluoric, hydrochloric, and boric acids for the subsequent determination of arsenic, barium, beryllium, cadmium, chromium, lead, mercury, silver, thallium, and tin.[120] Recoveries from oil-based reference materials ranged from 93% for barium to 100% for chromium.

3.2 CONCENTRATION AND SEPARATION METHODS

Liquid samples as well as the solutions resulting from the decomposition of solid samples sometimes require further treatment to raise the levels of analyte to the optimal concentration ranges for quantification by atomic spectrometry or to remove interfering materials from the solutions containing the analyte. There are four common procedures for such treatments.

1. *Evaporation of Solvent.* Although straightforward, this approach is vulnerable to losses by volatilization, and it increases both the concentration of analyte and interfering materials.
2. *Coprecipitation.* Coprecipitation of analyte followed by redissolution of the precipitate and analyte is a procedure that yields both increased concentration and a better defined matrix.
3. *Solvent extraction.* Extraction of analyte yields both increased concentration and a better defined matrix. Atomic absorption signals are often enhanced by organic solvents.
4. *Ion exchange chromatography.* This technique gives significant increases in

concentration of analyte and efficiently removes some interfering species, especially when chelating resins are used.

3.2.1 Evaporation of Solvent

Although the evaporation of solvent is an obvious approach to concentrating analytes, it has two major disadvantages. If the original sample has a high total dissolved solids (TDS) content, the TDS of the concentrated sample may become high enough to cause interferences in the measurement of atomic absorption/emission signals. It is also possible to experience losses of analyte through precipitation or coprecipitation if the solubility of some component of the system is exceeded during the evaporation. It is equally possible to experience losses by volatilization during evaporation. Of lesser significance is the slowness of the process: several hours is required to achieve a 10-fold concentration increase by evaporating a one-liter sample to 100 mL.

Both Method 200.2[58] and Method 200.7[121] make use of evaporation for the concentration of water samples prior to quantification by atomic spectrometry. These methods achieve, respectively, two- and fourfold concentration enhancements after hot plate or steam bath evaporation of water samples. Using the latter method, recoveries of samples spiked at half the maximum contaminant levels (MCLs) with arsenic, barium, cadmium, chromium, copper, iron, lead, manganese, silver, and zinc were 100 ± 5% for replicate determinations in tap water and groundwater matrices.

3.2.2 Coprecipitation

Coprecipitation offers 10- to 50-fold concentration increases. This technique, however, has several disadvantages: the precipitation process is lengthy and tedious; the precipitating agent must be of high purity to avoid contamination; and if the carrier precipitate (scavenging agent) cannot be destroyed during the redissolving process, the final solution will have a high TDS and will be subject to the same interferences encountered when solutions are concentrated by evaporation of solvent.

EPA Method 7195 made use of coprecipitation with lead sulfate from a pH 3.5 acetic acid medium for the separation of hexavalent chromium from trivalent chromium prior to the quantification of the former by atomic absorption spectrometry.[122] Similarly, Hudnik and Gomiscek[123] have used coprecipitation with iron(III) hydroxide to separate and concentrate part-per-billion arsenic and selenium from Slovene mineral waters prior to quantification by electrothermal atomization–atomic absorption spectrometry.

3.2.3 Solvent Extraction

Solvent extraction enjoys a favored position among separation techniques because of its speed, simplicity, and broad scope. Most extractions require only a few minutes to perform, often require apparatus no more complicated than a separatory fun-

TABLE 3.3 pH Ranges for the Formation of Metal–APDC Chelates and Their Extraction into MIBK[a]

	pH Ranges	
Metal Ion	Formation	Extraction
Antimony(III)	2–9	2–4
Arsenic(III)	2–6	2–6
Bismuth(III)	2–14	1–10
Cadmium(II)	2–14	1–10
Chromium(VI)	2–6	3–9
Cobalt(II)	2–14	1–10
Copper(II)	2–14	1–10
Iron(III)	2–14	2–5
Lead(II)	2–14	1–10
Manganese(II)	2–14	2–10
Mercury(II)	2–14	1–10
Nickel(II)	2–14	1–10
Selenium(VI)	2–9	3–6
Silver(I)	2–14	1–10
Tellerium(VI)	2–6	3–5
Thallium(I)	2–14	3–10
Zinc(II)	2–14	1–10

[a]Methyl isobutyl ketone, 4-methyl, pentanone-2.

nel, and can be applied both to isolate analyte and to remove interferants. Solubility in organic solvents is not a normal characteristic of typical inorganic salts. Their extraction from aqueous media most likely requires replacement of the water molecules hydrating the metal ion and neutralization of its charge. The formation of an extractable species frequently involves the formation of metal chelates. The Lewis acidity of the metal ion, the Lewis basicity of the polydentate ligand, and the pH of the system are important factors in the formation of such compounds. In the general case, the formation of the metal chelate, ML_2, is governed by a conditional formation constant:

$$K_f = \frac{ML_2}{(M^{2+})(L^-)^2}$$

The polydentate ligand, L^-, is the conjugate base of the weak acid HL. (L^-) is dependent upon both K_a of HL and the pH of the system. Pyrrolidine dithiocarbamic acid* forms chelate compounds with some two dozen metal ions. Table 3.3 lists the pH ranges for their formation and extraction.

*The ammonium salt ammonium pyrrolidine dithiocarbamate (APDC) is commonly used.

The formation of the metal chelate is governed by K_f. The extraction of the metal chelate depends on its relative solubility in the extracting organic phase and in the initial aqueous phase, that is,

$$D = \frac{(ML_2)_{org}}{(ML_2)_{aq}}$$

Efficient extraction demands favorable values for both K_f and D.

The extracting organic phase must possess the following characteristics: immiscibility with the aqueous phase, high solvency for the metal chelate, and chemical and physical properties compatible with atomization process. Methyl isobutyl ketone (MIBK) meets these requirements quite well: this efficient extractant for ammonium pyrrolidine dithiocarbamate chelates routinely allows 10-fold concentration increases. In addition, the atomization of the metal chelates is more efficient in MIBK than it is in aqueous media. This often leads to a two- to fivefold enhancement of the absorbance signal. Hence, the extraction of metal–APCD chelates from 100 mL of aqueous media into 10 mL of MIBK and subsequent aspiration of the organic phase often produces a 30- to 40-fold increase in the sensitivity of flame atomic absorption measurements. Interferences from monovalent cations and most anions are often eliminated with this chelation–extraction procedure.

EPA Method 7197 was based on the chelation of hexavalent chromium with APDC and subsequent extraction with MIBK prior to quantification by flame atomic absorption spectrometry.[124]

NIOSH Method 8003 makes use of chelation–extraction with APDC and MIBK for the determination of lead in blood and urine.[125] A 2.0 mL specimen of whole blood or filtered urine is treated with 0.8 mL of an aqueous solution containing 4 g APDC and 5 mL Triton X-100 per 200 mL, mixed with 2.00 mL of water-saturated MIBK, and centrifuged. The organic (upper) phase is aspirated into the air–acetylene flame for the quantification of lead by atomic absorption spectrometry. Lead recoveries from blood specimens spiked with 25–200 μg Pg/100 mL ranged from 94% to 106%, and lead recoveries from urine specimens spiked with 10 to 100 μg Pg/100 mL ranged from 95% to 106%. Average recoveries from both were 100%.

The EPA Special Extraction Procedure[126] also uses APDC–MIBK chelation–extraction for the isolation and concentration of cadmium(II), chromium(VI), copper(II), iron(III), lead(II), manganese(II), silver(I), and zinc(II). Alkali metal ions and those of the alkaline earths as well as the ions of aluminum(III), beryllium(II), and chromium(III) do not form complexes with APDC.

Richelmi and Baldi[127] have employed liquid ion exchangers (Amberlite LA-1 and LA-2) for the separation–extraction of hexavalent chromium from rat blood prior to atomic absorption spectrometry with electrothermal atomization.

The supercritical fluids method has been applied to the extraction of complexed metal ions from solid environmental samples.[128] Extraction from a cellulose matrix of various mercury species complexed with derivatives of diethyldithiocarbamate proceeded with efficiencies in excess of 90% using a 5% methanol modified carbon dioxide extractant.

3.2.4 Ion Exchange

Ion exchange techniques are frequently employed for the separation and/or concentration of analyte. Strong acid or strong base ion exchange resins are useful for separating cationic species such as calcium from anionic interferences such as phosphate. These resins have also been used to concentrate trace metals from dilute aqueous solutions. Strong acid and strong base resins, however, are of limited use in concentrating trace metal ions from seawater, urine, and blood plasma. Chelating resins, styrene–divinylbenzene copolymer containing iminodiacetate functional groups, have been used successfully for separating and concentrating transition metal ions from solutions of high salt concentrations.

Chelating resins differ from the conventional strong acid ion exchange resins in several respects: (1) their selectivity is a function of the iminodiacetate functional group rather than ionic size or charge; (2) the bond strength is almost 10 times greater than that of conventional strong acid resins; and (3) they have slower exchange kinetics than ion exchangers of other types. Chelating resin demonstrate very strong attraction for transition metal ions even from solutions high in salt concentration, hence have been used to separate them from alkali and alkaline earth metal ions as well as to concentrate the ions of cadmium, cobalt, copper, iron, lead, manganese, nickel, titanium, and vanadium.[129] EPA Method 200.10 employs an iminodiacetate resin for the on-line chelation preconcentration from seawater of cadmium, cobalt, copper, lead, nickel, uranium, and vanadium prior to their quantification by inductively coupled plasma–mass spectrometry.[130] In the course of preconcentration, separation from alkali and alkaline earth metal ions and from most anions was achieved. Spike recoveries were 94–104% at the 0.5 ppb level and 92–109% at the 10 ppb level.

Chelating resins containing functional groups other than the iminodiacetate have been applied successfully to the separation and concentration of trace metals prior to quantification by atomic spectrometry. Trace metal ions are separated and concentrated from urine by batch extraction with polydithiocarbamate chelating resin in NIOSH Method 8310. The procedure is as follows.[131] A 50.0 mL aliquot of acid-preserved urine is adjusted to pH 2.0 ± 0.1 with 5 M NaOH, treated with 60 ± 10 mg of polydithiocarbamate resin,* and agitated overnight. After this time the resin is recovered by filtration through a 0.8 μm cellulose ester membrane and retained. The filtrate is adjusted to pH 8.0 ± 0.1 with 5 M NaOH, treated with a second 60 ± 10 mg of polydithiocarbamate resin, and agitated overnight. After this time the resin is recovered by filtration through a 0.8 μm cellulose ester membrane and combined

**Preparation of polydithiocarbamate resin.* Solutions containing 72 g of polyethyleneimine (molar mass = 1800) in 1 liter of tetrahydrofuran and 28 g of polymethyleneolyphenyl isocyanate in 1 liter of tetrahydrofuran are poured simultaneously into a large flask so that the two streams mix before entering the flask. After standing for 12 hours with occasional mild agitation, the solid product is removed by filtration, and washed twice with methanol and once with distilled water. The product is added to 300 mL of carbon disulfide, 100 mL of ammonia solution, and 500 mL of 2 propanol and allowed to stand for 72 hours. The product is recovered by filtration, washed thrice with methanol, and allowed to air-dry. The dry product is ground and sieved. The 60/80 mesh fraction is retained for use.[131]

with the filter and resin from the first extraction. The combined filters and resins are either ashed or digested. In the former case, ashing proceeds in a low temperature oxygen plasma and is followed by dissolving the residues in 0.5 mL of 4:1 nitric acid–perchloric acid mixture. In the latter case, the filters and resins are digested with the acid mixture. The solutions resulting from either option are transferred to 5 mL volumetric flasks and brought to volume with distilled deionized water. The inductively coupled plasma–atomic emission spectrometry of aluminum, barium, cadmium, chromium, copper, iron, lead, manganese, molybdenum, nickel, platinum, silver, strontium, tin, titanium, and zinc separated and concentrated from spiked urine is presented in Table 3.4.

3.3 CONTAMINATION AND LOSS

Samples are subject to contamination or loss of analyte at any point in the collection–determination process. Therefore, special precautions are taken in cleaning the sampling devices and sample containers, in collecting the sample, and in ensuring maintenance of sample integrity through the addition of special preserving agents

TABLE 3.4 Inductively Coupled Plasma–Atomic Emission Spectrometry of Metals Recovered from Urine with Polydithiocarbamate Resin

Metal	Spike (μg/50 mL)	Coefficient of Variation (%)	Recovery (%)
Aluminum	20	17.2	100
Barium	0.4	41.6	80
Cadmium	1.0	23.5	100
Chromium	1.0	15.3	100
Copper	10	8.2	100
Iron	40	11.6	100
Lead	10	7.6	100
Manganese	10	113	85
Molybdenum	2.0	31.4	100
Nickel	2.0	102	80
Platinum	0.4	79.8	77
Silver	2.0	23.5	100
Strontium	4.0	49.0	100
Tin	2.0	41.2	100
Titanium	2.0	45.4	86
Zinc	200	17.4	100

Source: R. D. Hull, *ICP-AES Multi-Element Analysis of Industrial Hygiene Samples*, NIST Publication PB85-221414, National Institute of Standards and Technology, Gaithersburg, MD, 1985.

and the placement of limitations on holding times and conditions. Similar special precautions also apply to routine laboratory operations directed to sample preparation. Many of the problems involved in the sampling, storage, and preparation of biological materials for trace element determinations have been identified and reviewed.[66,132] Among the problems identified are contamination of the sample with analyte from extranal sources and loss of analyte from the sample during postcollection storage and treatment.

Some aspects of contamination and loss can be overcome with solid sampling. Solid sampling avoids contamination with impurities in the reagents and volatilization losses during vigorous digestion procedures. While direct introduction of solid samples into the atomizer has not yet been adopted by regulatory agencies, success with this approach has been reported in the literature. The results reported by Fujiwara et al.[133] show good agreement between the values for copper, iron, manganese, and zinc in silkworms and their eggs obtained by this technique and by neutron activation analysis. Chakrabarti et al.[134] obtained results within ±5% of the certified values for cadmium, cobalt, copper, iron, lead, and zinc in oyster tissue (NBS SRM 1566) by direct atomization from the solid state using the graphite furnace platform technique. Direct atomization from the solid state has also been applied to the determination of trace metals in alloys.[135] It would appear, however, that the small sample size places high demands on sample homogeneity. Introduction of the sample as a slurry has reduced the imprecisions associated with inhomogeneity.[136,137] Slurry sampling has been adopted to autosampling.[138,139] These developments may lead to real-time analysis.

3.3.1 Purity of Reagents

Because it is recognized that any chemicals added to prepare the sample for analysis are potential sources of contamination, special attention is given to the purity of reagents.

3.3.1.1 Acids Many of the sample preparation procedures described in Section 3.1.2.2 make use of digestion with nitric acid. Specifications for maximum impurity concentrations in two commercially available grades of "superior quality" nitric acid are listed in Table 3.5. Among the vendors of "superior quality" acids are J. T. Baker, Inc. (Phillipsburg, NJ), Fisher Scientific (Pittsburgh, PA), and Mallinckrodt (Paris, KY).

The following example will demonstrate the need for using ultra pure reagents in the preparation of biological tissues for trace metal determinations. Normal concentrations of chromium[140] and nickel[141] in human liver are 40 and 10 μg/kg dry mass, respectively. If a 200 mg specimen of human liver, which normally would contain 8 ng of chromium and 2 ng of nickel, were digested with 10 mL of the "Instra-Analyzed" nitric acid listed in Table 3.5, the chromium and nickel contributions from the acid would be, respectively, 500 and 30 ng. If the digestions were carried out with 10 mL of the Ultrex II nitric acid listed in Table 3.5, the chromium and nickel

TABLE 3.5 Impurities in Two Commercially Available Grades of Nitric Acid

	Nitric Acids	
Impurity	"Instra-Analyzed" (mg/L)[a]	Ultrex II (μg/L)[b]
Aluminum	0.03	<0.5
Arsenic		<0.1
Arsenic and antimony	0.005	
Barium	0.03	<0.02
Beryllium	0.01	<0.02
Bismuth	0.05	<0.02
Boron	0.005	<0.1
Cadmium	0.005	<0.02
Cesium		<0.02
Calcium	0.05	<0.5
Chromium	0.05	<0.1
Cobalt	0.01	<0.02
Copper	0.001	<0.1
Gallium	0.05	<0.02
Germanium	0.08	
Gold	0.05	<0.1
Iron	0.02	<0.5
Lead	0.0005	<0.1
Lithium	0.05	<0.02
Magnesium	0.007	<0.1
Manganese	0.002	<0.1
Mercury	0.0005	<0.5
Molybdenum	0.05	<0.1
Nickel	0.003	<0.1
Niobium	0.05	<0.02
Potassium	0.5	<0.5
Silicon	0.05	
Silver	0.01	<0.02
Sodium	0.5	<0.5
Strontium	0.005	<0.02
Tantalum	0.05	
Thallium	0.08	<0.02
Tin	0.005	<0.1
Titanium		<0.1
Vanadium	0.01	<0.02
Zinc	0.01	<0.1
Zirconium	0.01	<0.02

[a]Instra-analyzed Acids, product literature, J. T. Baker, Inc., Phillipsburg, NJ, 1990.

[b]Ultrex II Acids, product literature, J. T. Baker, Inc., Phillipsburg, NJ, 1994.

contributions from the acid would be no more than 1 ng of each. Clearly, only with the Ultrex II is the reagent blank reduced to a point that permits meaningful measurements to be made.

3.3.1.2 ***Water*** Successful trace metal analysis demands not only high purity acids but also high purity water. Several commercially available technologies are available for the production of reagent water. The American Society for Testing and Materials[142] has identified four categories of reagent water:

> TYPE I grade of reagent water shall be prepared by the distillation or other equal process, followed by polishing with a mixed bed of ion exchange material and a 0.2 μm membrane filter. Feedwater to the final polishing step must have a maximum conductivity of 20 μS/cm at 298 K (25°C).
>
> TYPE II grade of reagent water shall be prepared by the distillation using a still designed to produce a distillate having a conductivity of less than 1.0 μS/cm at 298 K (25°C). Ion exchange, distillation or reverse osmosis and organic adsorption may be required prior to distillation if the purity cannot be attained by a single distillation.
>
> TYPE III grade of reagent water shall be prepared by the distillation, ion exchange, reverse osmosis or a combination thereof followed by polishing with a 0.45 μm membrane filter.
>
> TYPE IV grade of reagent may be prepared by distillation, ion exchange, reverse osmosis, electrodialysis or a combination thereof.

The specifications for these four types of reagent-grade water are summarized in Table 3.6.

TABLE 3.6 ASTM Specifications for Reagent-Grade Water

Property	Type I	Type II	Type III	Type IV
Electrical conductivity, max, μS/cm at 298 K (25°C)	0.056	1.0	0.25	5.0
Electrical resistivity, min, MΩ·cm at 298 K (25°C)	18.0	1.0	4.0	0.2
pH at 298 K (25°C)[a]				5.0–8.0
Total organic carbon (TOC), max, μg/L	100	50	200	No limit
Sodium, max, μg/L	1	5	10	50
Chlorides, max, μg/L	1	5	10	50
Total silica, max, μg/L	3	3	500	No limit

[a]The measurement of pH in type I, II, and III reagent waters has been eliminated from the specification because these grades of water do not contain constituents in sufficient quantities to significantly alter the pH.

3.3.1.3 Other Reagents In addition to ultrapure acids and water, several vendors, Mallinckrodt in particular, supply a variety of specially prepared bases, buffers, complexing agents, and salts. Mallinckrodt Chemical offers hydrogen peroxide as Ar Select, the trace metal contents of which are at the level of parts per billion. The concentrations of many contaminants in the analogous Ar Select ammonia solution range from 45 ppb for calcium to less than 1 ppb for barium.

The reagent blank is important in identifying contamination of the sample with impurities from these reagents. When the reagent blank is less than 10% of the values measured for the samples, an appropriate correction factor may be applied to the results. If the reagent blank exceeds 10% of the values measured for the samples, purification of the reagents is required.

3.3.2 Contaminated Glassware A second source of contamination is improperly cleaned sample containers and laboratory glassware. The procedure described in Section 2.3.2.1 is adequate for preparing contamination-free sample containers and laboratory glassware. Once prepared, however, the vessels must be kept contamination-free. This can be accomplished by storage in sealed plastic bags or containers. Field blanks, trip blanks, and reagent blanks are useful in identifying contaminated glassware. It cannot be emphasized too strongly that the preparation and preservation of contamination-free glassware and plastic ware demands the same skill and commitment as the other aspects of regulatory compliance monitoring.

3.3.3 Losses During Storage

The adjustment to pH 2 with nitric acid for retarding losses of analyte from water samples by adsorption on container surfaces is described in Section 2.5.2. Complementing the work of Truitt and Weber[91] and of Subramanian et al.[90] cited in the earlier section are the demonstrations by Hoyle and Atkinson[92] and Das et al.[93] of the importance of carefully observing both the sample preservation and the holding time requirements specified in Section 2.5.2. Meanwhile, until on-site, real-time quantification becomes possible, losses during storage will remain a potential source of error in many aspects of regulatory compliance monitoring. The inclusion of spiked samples and other positive controls is helpful in the identification of losses during storage.

3.3.4 Losses During Preparation

Many of the potentials for losses by volatilization cited in the first edition of this book[143] have been rendered less acute with the introduction of closed-vessel, microwave-assisted digestion. The inclusion of standard/certified reference materials (SRMs/CRMs), spiked samples, and/or other positive controls is useful for identifying losses during preparation.

3.4 SAMPLE PREPARATION

No single preparation method is universally applicable to the diverse array of samples encountered in regulatory compliance monitoring. Associated with each method are both advantages and limitations for the quantification of constituent metals and metalloids. Laboratories whose data will be used to establish regulatory compliance must employ a common set of methods and procedures for sample preparation. It is important for these laboratories to recognize the need to utilize SRMs/CRMs for the validation of both methods and data.

4

QUALITY CONTROL/QUALITY ASSURANCE

A tremendous resource of environmental data is being generated annually by governmental, commercial, and industrial laboratories. These data are utilized for the assessment of the toxic and aesthetic qualities of drinking water, the determination of the acceptability of water and wastewater treatment, and also the development of water quality planning and management strategies. The importance of the decisions that are based on these data requires that only valid, defensible data of known quality be used.

Regulatory agencies have been placing more and more emphasis on quality assurance programs as a mechanism for identifying and documenting data quality. Quality assurance should not be confused with quality control. For purposes of this writing, quality control is defined as activities performed on a day-to-day basis to ensure that the data being generated are valid. Quality assurance is defined as an independent assessment of the monitoring process to ensure that the overall process (i.e., sample collection, sample analysis, quality control) that has been developed to generate valid data is functioning properly.

This chapter reviews the major points of a quality assurance program and discusses the important considerations that apply to any attempt to generate valid, defensible data.

4.1 LABORATORY CERTIFICATION

Many state and federal agencies operate laboratory certification programs as a type of quality assurance program that assesses and regulates the quality of environmental compliance data. The federal Safe Drinking Water Act regulates states wishing to assume primary enforcement responsibility for public water systems—that is, primacy—to establish and maintain a state program for the certification of laboratories conducting analytical measurements of drinking water contaminants.[144] No such federal requirement exists for other environmental regulatory programs. However, many state agencies identified the need to establish control over the quality of self-compliance monitoring data and have implemented laboratory certification programs covering other environmental matrices (e.g., wastewater, sludge, solid waste).[145]

Although the US EPA provides guidance to state agencies on the criteria to be used for certifying drinking water laboratories,[146] the proliferation of certification programs utilizing different criteria for certifying laboratories has led to the need to establish a national standard for all environmental laboratory certification programs to follow. Such an effort is under way through the formation of the National Environmental Laboratory Accreditation Conference (NELAC), which is sponsored by the Environmental Protection Agency in cooperation with the states, territories, and other federal agencies. NELAC, a voluntary association, was established for the purpose of fostering the "generation of environmental data of known quality through the development of national performance standards for environmental laboratories to be implemented by state and federal accrediting authorities in a consistent fashion nationwide."[147] The concept of NELAC is that the conference adopts national environmental laboratory certification standards to be applied later by all state and federal environmental laboratory accrediting agencies. It will take several years to implement this national venture and determine its success.

4.2 THE QUALITY ASSURANCE PROJECT PLAN (QAPP)

Another formal approach to quality assurance utilized by state and federal agencies features quality assurance project plans for environmental monitoring.[148] Such documents identify the data quality objectives of given projects, describe the steps that will be taken to meet the data quality objectives (sample collection and handling procedures, analytical methodologies to be used, quality control practices to be implemented, the data handling and validation procedures to be followed, etc.), and outline a process for auditing the entire data generating process to ensure that the monitoring system is functioning within the parameters established in the QAPP.

4.3 PLANNING

A major function of any quality assurance program is to provide a mechanism for communication between the designer of a monitoring project, the sample collection team, and the laboratory personnel. Communication between these three parties is essential if the project is to achieve its monitoring objective. That is, the sampling and laboratory personnel must have input to the monitoring project's design and identify the steps that must be taken to achieve the level of data quality necessary for intended usage of the data. In addition, any problems that are experienced with the sample collection and analysis process must be communicated back to the project designer and/or data interpreter.

When developing any monitoring project, it is important to identify and document:

Monitoring project objectives
Project personnel responsibilities and organization
Sampling procedures to be used
Analytical methodologies to be used
Instrumental calibration procedures
Quality control procedures to be followed
Procedure for data reduction and reporting
Quality assurance objectives and procedures

With this information in hand, the process for monitoring the quality of the data being generated and the responsibility for ensuring that the objectives of the quality assurance program are being met can be clearly defined and documented. Then throughout the monitoring project, the quality assurance coordinator, who is responsible for monitoring the quality of the data, should review all aspects of the monitoring process. As sampling or analytical problems arise, the quality assurance coordinator should qualify or reject the data the problem has impacted, review the problem, recommend corrective action, and monitor the situation closely to determine whether the corrective action has alleviated the problem. This process identifies invalid data and makes it possible to affirm that the data generated are of the quality required for the monitoring project objectives.

4.4 MONITORING PROJECT DESCRIPTION AND PERSONNEL OBJECTIVES

When generating data for any monitoring project, it is essential that all personnel involved understand the purpose of the project, their function in relation to the overall monitoring process, and the responsibility, as well as authority, placed on all participants. This may be accomplished by writing a brief project description and preparing an organizational chart of the overall project. In addition, the responsibility and authority of each individual must be clearly defined. If and when a problem arises during the project, this preparation will enable personnel to immediately notify the appropriate authority so that corrective action can be initiated and any data adversely affected can be qualified as "questionable results" or rejected.

4.5 ENVIRONMENTAL SAMPLING

4.5.1 Sample Collection

An integral part of any monitoring project, whether its focus is biological substances, wastewater, potable water, groundwater or surface water, is planning. Depending on the type of environmental system being monitored (e.g., stream, ambient air), the presurvey plan should first address three questions:

What is the objective of the monitoring project?
What type of sample is to be collected?
What method of sample collection is to be employed?

The objective of the monitoring project will dictate the type of sample to be collected (water, soil, air, etc.). However, as discussed in Chapter 2, there are many methods of sample collection. The major factor to be considered in the selection of method(s) of sample collection is that the objective of sampling is to collect a representative portion of the total environmental system being monitored. Other important factors are manpower resources available and sampling site location. From this information, the project planner should decide the frequency of sampling needed for the assessment, the specific sampling sites to be monitored, and the method of sample collection to be used (grab or composite).

4.5.2 Sample Labeling

All samples submitted to a laboratory for analysis should be properly labeled and accompanied by an analysis request form. The information put on labels and forms should be printed legibly. Cross-outs or erasures should not appear on the labels or forms: if mistakes are made, the label or form should be discarded and a new one filled out.

Sample labels, which are to be attached (by adhesives, elastic straps, etc.) to the sample container, should indicate:

Sample container number
Type of sample
Preservative method

The sample analysis request form should indicate:

Sample collector's name and organization
Name and identification number, if appropriate, of the laboratory performing the analysis
Sampling site location
Date and time of sample collection

The corresponding sample container number
Analytical test requested

The information, recorded on the label and form at the time of sample collection, should accompany the samples to the laboratory. When submitting samples to the laboratory, the sample collector should retain a copy of the analysis request form as a record.

4.5.3 Chain of Custody

All monitoring programs being conducted for compliance with state or federal regulations or programs involving data that may be used in a court of law should document and implement a chain of possession and custody of any sample material collected. This is necessary to assure and document that at all times the sample was under the control of one or a series of specific individuals (physical possession or a secured space) and could not have been damaged or altered in any way.

A chain of custody form, if used, should list at a minimum the following information:

Sample container number
Description of sample
Specific location of sample collection
Signature of sample collector
Date and time of sample collection

Once a sample has been submitted to a laboratory, the following information should be added to the chain-of-custody form (or the internal laboratory chain-of-custody form, if used):

Date and time of custody transfer to the laboratory (if the sample was collected by a person other than laboratory personnel)
Signature of the person accepting custody (if the sample was not collected by a member of the laboratory staff)
Date and time of initiation of analysis
Signature of person performing the analysis
Name of laboratory performing the analysis

Upon completion of the analysis, the chain-of-custody document(s) should be returned with the reported results to the sample collector.

4.5.4 Sample Handling and Preservation

As discussed in Chapter 2, careful consideration should be given to the type of sample container to be used for sample collection and transport to the laboratory. Glass or hard plastic[149] is recommended for trace metal samples.

Preservation techniques are designed to retard the chemical and biological changes that may occur in a sample after it has been removed from its source. In water samples, for example, metal cations may precipitate as hydroxides or form complexes with other constituents.

When sample preservation in matrices such as sludge or sediment samples is not easily accomplished, the sample should be refrigerated, transported immediately to the laboratory, and analyzed as soon as possible. Conditions under which sample preservation is difficult were outlined in Chapter 2.

When collecting potable water, wastewater, or other water-type samples, samples should be preserved with nitric acid to a pH below 2 immediately following sample collection. This may be accomplished by adding the acid (normally 3 mL of 1:1 nitric acid per liter of sample) to the sample bottle in the laboratory prior to sample collection or by adding the acid to the sample immediately following sample collection. The following is a recommended procedure for the latter case:

1. Add approximately 2 mL of 1:1 nitric acid to sample immediately following sample collection.
2. Replace stopper or cap on sample bottle and mix sample thoroughly by inverting bottle several times.
3. Remove sample bottle stopper and place a drop of sample from the stopper onto pH test paper.
4. Rinse the portion of the stopper exposed to the pH test paper with distilled or deionized water.
5. If pH is not less than 2, repeat steps 1 through 4.

Note: Care should be taken to follow these steps. Adding too much preservative ("overpreservation") will dilute the sample and yield inaccurate results.

When the samples have been properly collected and preserved, they should be transported to the laboratory and analyzed as soon as possible. Federal regulations require that properly preserved samples collected for drinking water analysis be analyzed within 6 months of the time of sample collection for trace metals (excluding mercury) and within 28 days for samples collected in glass or polyethylene containers for mercury analysis.[150] The United States Environmental Protection Agency also recommends that the foregoing holding times be applied to wastewater analysis.[151] The New Jersey Department of Environmental Protection requires that these holding times be applied to all samples being analyzed for compliance with the New Jersey Safe Drinking Water Act and the New Jersey Water Pollution Control Act.[152]

4.5.5 Field Logbooks

Each sample collector should have a bound daily logbook and take it into the field. At each sampling site, the sample collector should use the logbook to record:

Date and time of sampling
Specific sampling site
Sample container number
Field measurements (if conducted)
Weather conditions (if applicable to the survey)
Special comments (pertaining to the sample or sampling site, etc.)

All field logbooks should be neat, with entries written legibly, in ink. The sample collector should use the field logbook as a record of what samples were collected and submitted to a laboratory, keeping in mind that this logbook may be used in a court of law at some future date.

4.5.6 Quality Control

Field quality control addresses sampling equipment, field measurements, and sample collection procedures. A rule of thumb is that 5% of the field monitoring efforts should be directed toward quality control.

Procedures for calibration and maintenance of field instruments and sampling devices should be developed, implemented, and documented. These procedures would apply to stream flow monitoring devices, automatic samplers, and other automated equipment. The descriptive document should cover:

The cleaning procedure to be used on the equipment
The calibration procedure to be used on the equipment
Frequency of cleaning and calibration of the equipment

Also, if other field measurements are to be performed in conjunction with the survey (e.g., pH, dissolved oxygen), the procedures for calibrating and maintaining the associated instrumentation should be documented.

The following samples should be collected and analyzed for the same contaminants being monitored in the survey as part of the quality control check on the monitoring activities:

1. *Sample preservative blanks.* Sample preservatives, if used, may become contaminated in the field. Therefore, the same quantity of acid normally added to samples should be added to a sample bottle filled with laboratory pure water, and this sample should be submitted to the laboratory as a check on preservative contamination.
2. *Automatic sampler blanks.* If automatic samplers are used to collect samples, a quantity of laboratory pure water should be passed through the automatic sampler just after the device has been cleaned. This sample should also be submitted to the laboratory for analysis as a check on the efficiency of the cleaning procedure.

3. *Duplicate samples.* Duplicate samples should be collected at selected stations using two sets of sampling equipment at the same site or duplicate grab samples. These samples should be submitted to the laboratory for analysis as a check on the sampling equipment and on the precision of the technique.
4. *Split samples.* Two representative subsamples are removed from one collected sample and submitted to the laboratory for analysis. These data may be used as a check on the precision of the laboratory's analytical procedure.

4.6 LABORATORY ANALYSIS

4.6.1 Sample Receipt

When samples are submitted to a laboratory, laboratory personnel should check the integrity of the sample. That is, the samples should be checked for:

Proper identification (i.e., site of sample collection, collector's name)
Date and time of sample collection
Whether the sample is properly preserved
Whether sample volume is sufficient for the analysis requested
Proper sample container

It is important to check these items to avoid problems that may arise later during the analytical process. When checking the date and time of the sample collection, the analyst should determine whether accepting the material submitted will violate any regulated sample holding times (refer to Chapter 2). When checking the acid or base preservation of a sample, the sample receiver should analyze a portion with a pH meter to ascertain whether the sample was properly preserved, and document the finding. The sample receiver should also check the sample volume to determine whether there is enough sample present not only to conduct the analysis but also to meet the requirements for quality control sample volumes.

Upon checking the integrity of the sample, the sample receiver should decide on and document its acceptability. Samples improperly preserved, exceeding established holding times, or collected in improper sample containers should be rejected by the sample receiver. If the sample submitter insists that the sample be analyzed, then at a minimum, the data that are generated are qualified as "questionable results" and made available to researchers. The sample receiver should keep in mind that upon receiving the sample, the laboratory assumes full responsibility for that material.

4.6.2 Sample Analysis

The selection of the analytical method to be used for a particular analysis should be based on the monitoring objectives. If the sample has been collected as part of a state or federal compliance monitoring requirement, the method of sample analysis may have been designated by that agency (refer to Chapters 3, 4, and 7).

All analytical methods must be validated and the precision, accuracy, and method detection limit documented prior to routine usage. These steps apply to:

The setup of the procedure

The establishment of the method's calibration curve, its linearity, and the purity of reagents through the analysis of reagent blanks

The establishment of the method detection limit (MDL)

The establishment of percent recovery (%R) and relative standard deviation (RSD) of the method through the analysis of a quality control standard, for a minimum of 10 independent analyses

The analysis of a reference standard of the same sample matrix and documentation of the method's performance

Documentation of the analytical procedure

This information should be well documented and compared to the monitoring objective and/or the method's performance criteria, if available, to determine the acceptability of the method.

4.6.3 Routine Quality Control

As stated earlier, quality control is defined as the activities performed on a day-to-day basis to ensure that the data being generated are valid. This task should take up, at a minimum, 10% of the laboratory's efforts. This includes not only the conducting of spiked and replicate analyses, but also performing checks and tests on the instrumentation and reagents used in the course of the analytical process. Sections 4.6.3.1–4.6.3.3 are quality control checks specific to atomic absorption spectrometry that should be performed routinely, and documented.

4.6.3.1 Instrument Quality Control

Atomic Absorption Spectrophotometer. After the analyst has followed the manufacturer's instructions for adjusting and calibrating the atomic absorption spectrophotometer, the instrument's maximum obtainable sensitivity should be checked for the particular element at hand by analyzing a specific standard. The following information should be recorded in a bound notebook: the standard's concentration and absorption reading, the date of check, and the name of the analyst. This process should be repeated each time the instrument is set up for a particular analysis, to verify the instrument's stability.

Conductivity Meter. The conductivity meters used for checking the purity of the laboratory water (as discussed below) should be checked annually. Meters equipped with conductivity cells having platinum electrodes should be checked over the range of interest using at least five concentrations of a standard potassium chloride solution. Meters not equipped with platinum electrodes should be checked against a conductivity meter equipped with platinum electrodes. The following data should be recorded in a bound notebook: the raw data, the cell constant, the correction factor (if needed), the comparison results, the date of check, and the analyst's name.

Analytical Balance. If an analytical balance is used in the preparation of calibration standards, the device should be checked and adjusted annually by a service person employed by the laboratory, or by a balance consultant. The accuracy of the balance should be checked once a month using at least two class "S" weights (one in the 5–50 g range and one in the 10–500 mg range). The weights used in the monthly check, the weight detected to the nearest 0.1 mg, the dates on which the checks were performed, the analyst's name, and other pertinent information should be recorded in a bound notebook.

pH Meter. The pH meter used to check the proper acidification of the sample at the time of sample receipt must be calibrated each day of use. The calibration should be conducted with two standard pH buffers bracketing the value to be measured. After calibration, a standard buffer with pH within the calibration range should be measured without any control adjustments to check the calibration. All calibration and check data should be recorded in a bound notebook, signed and dated by the analyst. When the pH meter is in use for a period longer than 3 hours, the pH of the third buffer should be checked once every 3 hours. If the pH differs by more than ±0.2 pH unit from the standard buffer, the meter should be recalibrated.[153]

4.6.3.2 Reagent Quality Control

Laboratory Pure Water. To avoid the introduction of contaminants into the analyses by way of the laboratory pure water that may be used (refer to Section 3.3.1), the purity of the water should be checked daily by means of a conductivity meter. The conductivity reading, the date of check, and the name of the analyst should be recorded in a bound notebook. Preferably, laboratory pure water should have a conductivity of less than 1 μmho/cm (1 μS/cm).

Reagents and Chemicals. Spectroquality chemicals should be used for trace metal analysis, although sometimes reagent-grade quality may be satisfactory. A reagent blank, as discussed below, should be analyzed to determine whether a particular reagent contains a contaminant or chemical that may interfere with a particular analysis. The following practices should be followed:

1. All chemicals, solutions, and standards should be dated upon receipt.
2. All solutions should be properly labeled to identify the compound, state its concentration and give the date of preparation and the name of the analyst who prepared the solution.

3. Stock and working solutions should be checked regularly for signs of discoloration, formation of precipitates, and concentration change due to evaporation.

Standards. Standards are an integral part of the quality assurance/quality control program. In addition to their uses in calibration and recovery (spiking), standards serve in the evaluation of precision and accuracy. Hence, standards are available for both calibration standards and reference standards.

CALIBRATION STANDARDS. Calibration standards may be prepared in-house or purchased from commercial sources. Standards for atomic absorption spectrometry are usually prepared at concentrations of 1000 mg/L ± 1% in aqueous nitric acid matrix. Some vendors supply standards in nonaqueous media. Standards for inductively coupled plasma–atomic emission spectrometry and for inductively coupled plasma–mass spectrometry are available as single-element concentrates containing either 1000 or 10,000 mg/L ±0.3% of analyte, or as multielement solutions in which the concentrations of analytes are adjusted to match those expected to be encountered in the samples. In addition to serving as the mixed calibration standards, multielement solutions also serve as the interference check solutions in the US EPA Methods 200.7 and 6010. The regulatory agencies often require calibration standards to be traceable to NIST. Some commercial sources of calibration standards are listed in Table 4.1.

REFERENCE STANDARDS. Like calibration standards, reference standards may be prepared in-house or purchased from commercial sources. The International Atomic Energy Agency[154,155] (IAEA) has catalogued the standard and certified reference materials (SRMs and CRMs) available from national and international agencies. This worldwide inventory lists sources from which samples of biological, environmental, and geological materials, commercial products, and domestic and industrial wastes can be obtained. The elemental compositions of these samples were established by intra- and interlaboratory analyses and statistical evaluations. Reference materials are available from commercial sources also. Some of these sources are listed in Table 4.2.

The National Institute of Standards and Technology is an international leader in the preparation and standardization of reference materials. As such, only NIST may issue SRMs. Other agencies such as the National Research Council of Canada (NRC) provide CRMs. NIST Special Publications 829[156] and 260-100[157] provide valuable information on how SRMs can be used to evaluate accuracy and precision.

4.6.3.3 *Analytical Quality Control*

Reagent Blanks. Each individual reagent used in the analytical procedure should be tested to determine whether it causes any interference with the analysis. The conditions for handling and analyzing the blank should be identical to those used in the

TABLE 4.1 Some Commercial Sources of Calibration Standards for Atomic Spectrometry

Vendor	Products
J. T. Baker, Inc. Phillipsburg, NJ	Single-element standards for AAS; single- and multielement standards for ICP-AES and ICP-MS; performance check and interference check solutions; nonaqueous standards
EM Science Gibbstown, NJ	Single-element standards for AAS; single- and multielement standards for ICP-AES and ICP-MS; performance check and interference check solutions; nonaqueous standards
Environmental Express Mount Pleasant, SC	Single-element standards for AAS; single- and multielement standards for ICP-AES and ICP-MS; performance check and interference check solutions
Environmental Resources Arvada, CO	Single-element standards for AAS; performance check and interference check solutions
High-Purity Standards Charleston, SC	Metal and compounds; single-element standards for AAS; single- and multielement standards for ICP-AES and ICP-MS; performance check and interference check solutions; nonaqueous standards
Inorganic Ventures Lakewood, NJ	Metal and compounds; single-element standards for AAS; single- and multielement standards for ICP-AES and ICP-MS; performance check and interference check solutions; nonaqueous standards
Promochem, GmbH Wesel, Germany	Aqueous and nonaqueous single-element standards
Radian Corporation Austin, TX	Single-element standards for ICP-AES and ICP-MS
SPEX Industries Edison, NJ	Single-element standards for AAS; single- and multielement standards for ICP-AES and ICP-MS; performance check and interference check solutions; nonaqueous standards
Ultra Scientific North Kingston, RI	Single-element standards for AAS, ICP-AES, and ICP-MS

analysis. The reagent blank should be analyzed with each new reagent and data generated from the reagent blank should be documented.

Method Blank. The method blank is analyzed to determine whether the cumulative effect of the reagents causes interference with the analysis. The method blank should consist only of laboratory pure water and the reagents used in the analysis. The method blank is handled and analyzed in the same manner as standards and samples. A method blank should be analyzed each time an analysis is conducted.

TABLE 4.2 Some Commercial Sources of Reference Standards for Atomic Spectrometry

Vendor	Products
Brammer Standard Houston, TX	Glasses, ceramics, alloys, ores, rocks, slag, soil from international agencies
EM Science Gibbstown, NJ	Simulated reference materials for drinking water, groundwater, and TCLP extracts
Environmental Express Mount Pleasant, SC	Simulated reference materials for drinking water, groundwater, wastewater, and TCLP extracts
Environmental Resource Arvada, CO	Simulated rainwater, drinking water, and wastewater
High-Purity Standards Charleston, SC	Wastewater and TCLP extract; solutions from digested fish, sediment, and soil; sandy and loamy soils; spiked cellulose air filters
Inorganic Ventures Lakewood, NJ	Simulated reference materials for drinking water, groundwater, wastewater, and TCLP extracts; soils, ashes, paint chips, water filter media, and sludges from wastewater treatment and electroplating
International Atomic Energy Agency, Vienna, Austria	Many authentic biological, environmental, and geological certified reference materials
National Institute of Standards and Technology, Gaithersburg, MD	Many authentic biological, environmental, geological, and industrial standard reference materials
National Research Council Halifax, Nova Scotia, Canada	Certified reference materials prepared from authentic marine waters and sediments and tissues of marine animals
Promochem, GmbH Wesel, Germany	CRMs for plant and animal tissue; oil, coal, coke, and ash; air, water, and soil; and clinical specimens
SPEX Industries Edison, NJ	Simulated reference materials for drinking water, groundwater, wastewater, and TCLP extracts
Ultra Scientific North Kingston, RI	Simulated reference materials for drinking water and wastewater

Calibration Curve Linearity. A minimum of one method blank and three standards should be used in the establishment of a calibration curve. The linearity of the curve should checked by calculating its correlation coefficient, which should be greater than 0.995. The calibration curve should be verified by analyzing a calibration check standard after every 20 samples. The data used in establishing the calibration curve must be recorded and signed by the analyst. This record should indicate the date of calibration, as well as the identification and concentration of each standard.

Method Detection Limit (MDL). The analytical limit of detection should be documented. There are several approaches to documenting the limitations of the analytical method's ability to detect and/or quantitate a certain analyte (e.g., limit of detection, limit of quantitation).[158] Most environmental agencies require that the MDL be established annually for each analytical method in accordance with the procedure given in the Code of Federal Regulations (40 CFR Part 136, Appendix B).

Laboratory Control Sample. Laboratory control samples should be analyzed each time an analysis is performed. The resultant data are used to determine whether the analytical process is within acceptable control limits.

For each parameter being analyzed, the standard deviation (s) should be calculated and documented; the primary standard used for this task should be of a different source from the primary standard used to generate the calibration curve. This information is then plotted on a traditional quality control chart establishing upper and lower control limits based upon $3s$. The quality control standard used to establish "s" is then reanalyzed with subsequent analytical runs and the data plotted on the quality control chart to determine whether the analysis is in control.[50] To this end, a specific concentration of a standard may be analyzed and the analytical result documented. After 20 determinations have been made, the standard deviation may be calculated using the following equation:

$$s = \left[\frac{\Sigma(x - \bar{x})}{n - 1} \right]^{1/2}$$

where Σ is the summation of values, x is the observed value, $\bar{x}$ is the mean or average value, and n is the number of observations. *Example:* Over a period of 2 months, a 5 ppm copper standard was analyzed 20 times and the standard deviation was calculated (Table 4.3).

Once calculated, the standard deviation can be expressed by referring to a quality control chart to determine whether an analyst is in control of the analyses. The control chart in Figure 4.1 is a plot of the standard's mean value ($\bar{x}$), observed value, upper control limit (UCL), and the lower control limit (LCL). Typically, the UCL and LCL are based on 2 standard deviations from the standard's mean value. When a control chart has been generated for a particular analysis, the standard used to calculate the standard deviation is analyzed with each future analytical run. The value determined for that standard's analysis is then plotted on the control chart. If the value determined exceeds either the UCL or the LCL, the analysis is termed "out of control" and the analyst begins checking the procedure for the error. Upon locating and correcting the problem (e.g., instrument failure, need to replace reagents), the analyst reruns the standard to verify that the problem has been corrected. Figure 4.2 is an example of a control chart using the data listed in Table 4.3. Quality control charts can be a simple graphical means for analysts to document and review quality control data.

TABLE 4.3 Standard Deviation Calculation

Observation	Detected Value	$x - \bar{x}$	$(x - \bar{x})^2$
1	4.89	−0.10	0.01
2	5.01	0.02	0.0004
3	4.98	−0.01	0.01
4	4.95	−0.04	0.0016
5	5.03	0.04	0.0016
6	5.03	0.04	0.0016
7	5.14	0.15	0.0225
8	5.03	0.04	0.0016
9	4.95	−0.04	0.0016
10	5.01	0.02	0.0004
11	5.02	0.03	0.0009
12	4.98	−0.01	0.0001
13	4.97	−0.02	0.0004
14	5.04	0.05	0.0025
15	5.05	0.06	0.0036
16	5.01	0.02	0.0036
17	4.93	−0.06	0.0036
18	4.90	−0.09	0.0081
19	4.98	−0.01	0.0001
20	4.96	−0.03	0.0009
	99.86		

$$\frac{99.86}{20} = 4.99 \qquad \Sigma(x - \bar{x})^2 = 0.0719$$

$$s = \left[\frac{\Sigma(x - \bar{x})}{n - 1}\right]^{1/2} = 0.06$$

Spiked Analyses. A minimum of 5% of all samples being analyzed should be analyzed as spiked samples. Spiked analyses are performed by splitting a sample into replicates and adding to one of the replicates a known amount of the contaminant being tested for. The amount of the added contaminate should be should be approximately the same as the amount present in the unspiked sample. Both samples should then be analyzed and the percentage recovery of the spike may be addressed as follows:

$$R = \frac{100(F - I)}{A}$$

where F is the analytical result of the spiked sample, I is the result before spiking of the sample, and A is the amount of contaminant added to the sample. This information should be documented.

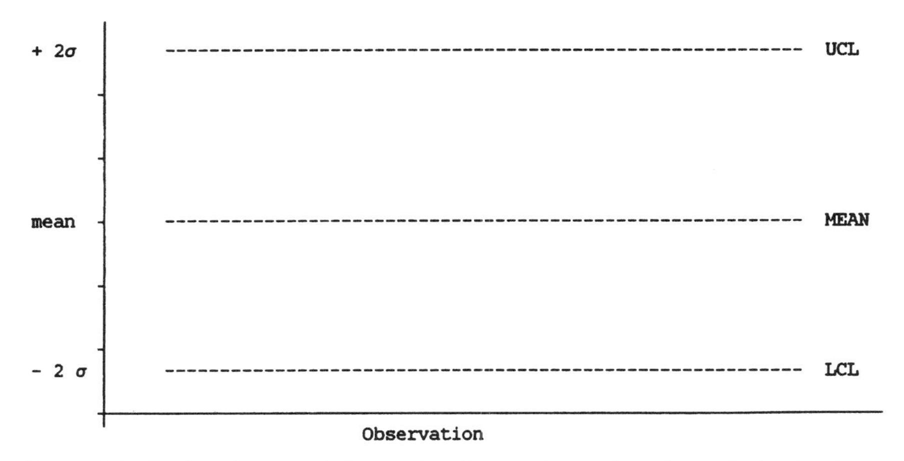

Figure 4.1 Basics of a control chart: unit values are inserted on the vertical axis; observation numbers are arrayed horizontally.

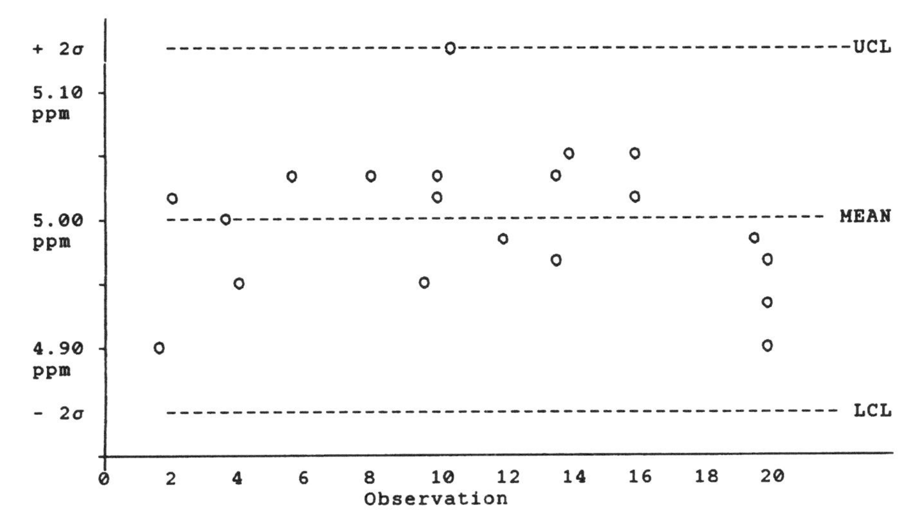

Figure 4.2 Example of a control chart: $\bar{x} = 4.99$; $s = 0.06$; UCL = 4.99 + 2s = 5.11; LCL = 4.99 – 2s = 4.87.

Duplicate Analyses. A minimum of 5% of all samples being analyzed should be tested in duplicate. Duplicate samples are prepared by dividing a homogeneous sample into separate parts so that each part is also homogeneous and representative of the original sample. The data obtained from the duplicate analyses should be used to document the precision of that particular method.

Duplicate Spiked Analyses. In an effort to avoid generating data of limited usage (e.g., analyte in question less that detection limit for particular sample selected for duplicate analyses), many regulatory agencies now require that a duplicate spiked analysis be conducted in lieu of the independent spike and duplicate analysis.

4.6.4 Data Handling and Documentation

When the analysis of the sample has been completed, the following should be recorded in the analyst's notebook:

Date and time of analysis
Analytical method employed
Raw data of standards
Quality control raw data
Sample number, raw data, and final result
Calculations used to determine final result
Analyst's signature

A review process should also be incorporated into the laboratory procedure. That is, a second analyst should review the first analyst's quality control data, raw data, and calculations to verify the reported results. Upon completion, the second analyst should also initial the notebook. Then, after the document has been checked for any transcription errors carried through the reporting process, the data may be submitted to the sampling agency.

All laboratory data (raw analytical data, quality control data, sample reports) should be well documented, and these records should be maintained as a permanent record for many years. If at any time the validity or nature of the results is questioned, these records will be needed for purposes of verification.

When computers or laboratory automation systems are used in the capture, transfer, processing, reporting, or storage of analytical data, the computer software and system operating procedures should be well documented. In addition, the EPA has drafted guidelines for good automated laboratory practices that set minimum standards for establishing and documenting laboratory automation.[159]

4.6.5 Chain of Custody

As noted earlier, the data generated are sometimes used in a court of law. If this contingency may reasonably be anticipated, the laboratory should establish a set of protocols designed to ensure and document the integrity of samples and to permit verification of their custodial history.

The laboratory's chain of custody procedures should begin with the receipt of the samples and continue through the analytical process. The chain-of-custody form used should indicate at a minimum the following:

Date and time the laboratory received custody of the sample
Person accepting custody of the sample
Whether the sample was received preserved or unpreserved
Date and time of analysis
Person or persons who performed the analysis
Type of analysis performed and analytical method employed

The completed chain-of-custody form should be attached to the analytical report and forwarded to the sampling agency.

4.7 QUALITY ASSURANCE DOCUMENTATION

This chapter has stressed two main points that should be addressed whenever a monitoring project is being designed, mainly (1) defining each step of the monitoring process and (2) documenting all steps prior to and during the process. The amount of work involved in performing these tasks may be minimized through the use of standard operating procedure (SOP) manuals. A field SOP and a laboratory SOP should be developed and maintained through the monitoring program. These manuals document the routine procedures practiced and may serve as guides or instruction manuals to all personnel.

When the SOPs have been established, routine quality control record-keeping practices (e.g., notebook formats for each analyst and instrument) should be developed. When a routine recording procedure is in place, the amount of time devoted to maintaining the records is kept to a minimum and the only major costs associated with quality control entail the quality control analysis, review of the quality control data, and implementation of the data when a problem arises—usually 10–15% of the overall monitoring program costs. This proportion is considered reasonable compared to the consequences that may ensue if decisions are based on invalid data.

In addition to the laboratory SOPs, the laboratory should develop and maintain laboratory methods manuals, to describe, in detail, the methods for all in-house analyses performed. These manuals provide a historical record of the exact methods employed by the laboratory and indicate the specific options taken (if provided in the method) or deviations made (if legally allowed) when conducting any particular analysis.

4.8 QUALITY ASSURANCE PROGRAM

Each laboratory should have a quality assurance program that:

1. Conducts systems audits to ensure that the standards and procedures established by the laboratory are being met.

2. Tests the laboratory's analytical performance through the use of "blind" laboratory performance evaluation samples.
3. Monitors the overall reliability of the analytical results.

The systems audit should be an independent review of the analytical process to determine whether the procedures set out in the quality assurance project plan and/or laboratory SOP manual are being followed and whether the desired data quality objectives are being met. Blind performance evaluation samples are used in conjunction with standard reference materials, blinded or nonblinded, to assess the analytical accuracy of a given routine.

In smaller laboratories, the analyst may be the individual responsible for quality assurance, and participation in a laboratory certification program may be sufficient for assuring that the laboratory is generating accurate and defensible data. Larger and more diverse laboratories may find it necessary to assign to one quality assurance officer the responsibility for carrying out the activities associated with assessing the analytical system. In either case, the quality assurance program should be designed to document acceptable performance, identify problems that may occur, initiate corrective action, and test the analytical system to verify its importance. The end result is the assurance of all analytical results produced by the laboratory.

5

METHODS FOR COMPLIANCE AIR QUALITY MONITORING

The methods described in this chapter are applicable to compliance monitoring of air quality in the workplace. These methods are described in more detail in the *NIOSH Manual of Analytical Methods,* 4th ed.[79] They are a necessary part of the standards that define limits for environmental exposures to toxic substances, and they have been incorporated into the criteria documents produced by the National Institute for Occupational Safety and Health (NIOSH) for recommending standards to the Occupational Safety and Health Administration (OSHA). This chapter also describes the US EPA methods for monitoring emissions of some metals from various stationary sources.

5.1 GENERAL PROCEDURES FOR METALS

Trace metals in airborne particulate matter are frequently identified and quantified by atomic spectrometry. Samples are collected by drawing ambient air through a filter at a known rate for a measured time. The particulate matter is dissolved in acid, and the concentrations of trace metals in the resulting solutions are determined either by inductively coupled plasma–atomic emission spectrometry or by atomic absorption spectrometry. Rigorous programs of quality control are required to validate the data.

5.1.1 General Procedure for Metals; NIOSH Method 7300

NIOSH Method 7300 is a procedure for the collection, dissolution, and determination of trace metals in industrial and ambient airborne material. Samples are collected on membrane filters and dissolved with a nitric acid–perchloric acid mixture.

The aluminum, arsenic, beryllium, cadmium, calcium, chromium, cobalt, copper, iron, lead, lithium, magnesium, manganese, molybdenum, nickel, phosphorus, platinum, selenium, silver, sodium, tellurium, thallium, titanium, vanadium, yttrium, zinc, and zirconium contents are quantified by inductively coupled plasma–atomic emission spectrometry (ICP AES).

For personnel sampling, the worker is fitted with a 37 mm diameter, 0.8 μm pore size cellulose ester membrane filter in a lapel holder connected to a calibrated pump for the collection of a 200–2000 L sample, the actual volume of which depends on the permissible exposure limits (PELs) of the analytes. The flow rate, ambient temperature, and ambient pressure should be recorded at the beginning and again at the end of the sampling period. At the end of the sampling period, the filter holder is removed from the worker's lapel and the inlet and outlet are capped with the original plugs. Although the sample is stable, the sealed filter holder should be returned to the laboratory without delay. Cellulose ester membrane filter field blanks should be included with the shipment.

In the laboratory the filters are carefully removed from the holders, transferred to clean, labeled 50 mL beakers, and treated with 5 mL of 4:1 (v/v) nitric acid–perchloric acid mixture. Spiked cellulose ester membrane filters should be included as positive controls, and additional beakers containing 5 mL of 4:1 (v/v) nitric acid–perchloric acid mixture should be prepared as reagent blanks. The beakers are covered with watch glasses and allowed to stand for 30 minutes at room temperature. The beakers containing the sample filters, the field blanks, and the reagent blanks are then heated on a hot plate at 120°C until the filters have dissolved and the volumes of the contents are reduced to approximately 0.5 mL. Additional 2 mL increments of 4:1 (v/v) nitric acid–perchloric acid mixture are added, and heating is resumed until clear solutions are obtained. The contents of the beakers are then treated with 2 mL of 1:12 dilution of the 4:1 (v/v) nitric acid–perchloric acid mixture in water, cooled, and quantitatively transferred to 10 mL volumetric flasks. The contents of the flasks are brought to volume with a 1:12 dilution f the 4:1 (v/v) nitric acid–perchloric acid mixture in water.

Quantification of aluminum, arsenic, beryllium, calcium, cadmium, chromium, cobalt, copper, iron, lead, lithium, magnesium, manganese, molybdenum, nickel, phosphorus, platinum, selenium, silver, sodium, tellurium, thallium, titanium, vanadium, yttrium, zinc, and zirconium in the solutions is by ICP-AES relative to standards prepared in aqueous 4:1 (v/v) nitric acid–perchloric acid mixture. With the following exceptions, measurements are made at the wavelengths listed in Table 1.3: arsenic at 193.7 nm, calcium at 315.9 nm, cobalt at 231.2 nm, chromium at 205.6 nm, molybdenum at 281.6 nm, selenium at 190.6, titanium at 334.9 nm, and vanadium at 310.2 nm.

Although some compounds of aluminum, beryllium, chromium, cobalt, lead, lithium, manganese, molybdenum, platinum, and zirconium may not be dissolved completely by the 4:1 (v/v) nitric acid–perchloric acid mixture, recoveries from cellulose ester membrane filters spiked with 2.5 and 1000 μg of the elements listed ranged from 75 to 105%. Precisions of triplicate measurements were within ±10%.

5.1.2 General Procedure for Metals; ASTM Designation D 4185

Designation D 4185[46] describes the collection, dissolution, and determination of two dozen trace metals in workplace atmospheres.

Samples are collected on 0.8 μm pore size cellulose ester or cellulose nitrate membrane filters at 2 L/min. The filters with samples and those from the field blanks are transferred to 125 mL beakers, treated with concentrated nitric acid, covered with watch glasses, and placed on a 140°C hot plate for 30 minutes. This sequence should solubilize most components of most samples, but some compounds of aluminum, chromium, and cobalt may require additional treatment with mixed acids. When dissolution is complete, the contents of the beakers are transferred to 10 mL volumetric flasks and brought to volume. Further dilutions and treatments with releasing agents, ionization buffers, and so on are made as needed.

The techniques for flame atomic absorption spectrometry from Table 1.1 are applicable to the determination of aluminum, barium, bismuth, cadmium, calcium, chromium, cobalt, copper, indium, iron, lead, lithium, magnesium, manganese, nickel, potassium, rubidium, silver, sodium, strontium, titanium, vanadium, and zinc. The atomic absorptions of bismuth, indium, lithium, rubidium, and strontium are measured at 223.1, 303.9, 670.8, 680.0, and 460.7 nm, respectively, in an oxidizing air–acetylene flame. A 1000 ppm cesium (as CsCl) ionization buffer is recommended for the last three analytes.

5.2 PROCEDURES FOR ALUMINUM; NIOSH METHOD 7013

In the NIOSH method for the determination of aluminum in airborne particulates, samples are collected on membrane filters and dissolved in nitric acid. The aluminum concentrations of the resulting solutions are determined by flame atomic absorption spectrometry.

Particulates from a 200 L air sample are collected using the filter and procedure described in Section 5.1.1. The filters with the samples, the field blanks, and the spiked filters are transferred to clean beakers. To each of these plus each of those for the reagent blanks is added 6 mL of concentrated nitric acid. The beakers are covered with watch glasses and heated on a 140°C hot plate. When clear solutions have been obtained, the watch glasses are removed from the beakers and the contents evaporated to approximately 0.5 mL. The contents of the beakers are treated with 4 mL of 10% nitric acid, heated briefly to assure complete solution of the digested material, and transferred to 10 mL volumetric flasks. After the addition of 0.2 mL of 75 mg/mL cesium nitrate ionization suppressor, the contents of the flasks are brought to volume with 10% nitric acid.

Note: Alumina (Al_2O_3) will not be dissolved by this procedure. Lithium borate fusion is necessary to dissolve alumina. The fusion procedures described in Section 3.1.2.1 are applicable.

The aluminum concentrations of the solutions are determined by atomic absorption spectrometry in the nitrous oxide–acetylene flame at 309.3 nm relative to aluminum standards in 10% nitric acid containing 1000 ppm cesium ionization suppressor.

5.3 PROCEDURES FOR ARSENIC

The vast number of arsenic compounds having toxicological and environmental significance has made necessary the development of methodologies for their identification and quantification. Some toxicological and environmental aspects of arsenic chemistry have been described by Nriagu,[160] and some of the methodologies for the identification and quantification of airborne arsenic compounds are described in Sections 5.3.1–5.3.5.

5.3.1 Determination of Organoarsenic; NIOSH Method 5022

NIOSH Method 5022 is applicable to the identification and quantification of methylarsonic acid, dimethylarsenic acid, and *p*-aminophenylarsonic acid in airborne particulate matter. Samples are collected on poly(tetrafluoroethylene) membrane filters, and the arsonic/arsenic acids are solubilized in a borate buffer. The arsonic/arsenic acids are resolved by ion chromatography and quantified by on-line atomic absorption spectrometry of the gaseous hydride.

Airborne particulates from 500 L samples are collected on polyethylene-backed, 1 μm pour size PTFE membrane filters (Millipore type FA or equivalent) with backup pads at flow rates of 2 L/min. The cassettes containing the filters with the samples and those serving as field blanks are plugged and returned to the laboratory for evaluation.

The filters with the samples, the field blanks, and spiked filters are transferred to 50 mL beakers and 25 mL of an extracting solution (2.4 mM $NaHCO_3$, 1.9 mM Na_2CO_3, and 1.0 mM $Na_4B_2O_7 \cdot 10H_2O$) is added to each. Reagent blanks are prepared also. The beakers are covered with watch glasses, and their contents are subjected to ultrasonic agitation for 30 minutes in a water bath. The extracts should be stored at 4°C if the identification/quantification of the arsonic/arsenic acids is delayed.

A 2.5 mL aliquot of extract in introduced via the injection loop into the ion chromatograph containing two 3 × 150 mm anion columns. The chromatographic separation is conducted with the bicarbonate–carbonate–borate extracting solution just described at 2.5 mL/min and 500 psi. The suppressor column and the detector are bypassed, and the effluent from the anion exchange column is routed to the arsine generator through microbore PTFE tubing.

The arsine generator consists of a proportionating pump to first mix the column effluent with 0.8 mL/min saturated potassium persulfate in 15% (v/v) hydrochloric acid solution and then with 2 mL/min 1% (m/v) sodium borohydride in 2% (m/v) potassium hydroxide solution. The reaction products are carrier to a gas–liquid separator with argon flowing at 300 mL/min, where the liquid phase is drained to waste and the gaseous phase is introduced into the quartz furnace for atomic absorption spectrometry at 193.7 nm.

Elution times are approximately 1 minute for dimethylarsenic acid, 2 minutes for methylarsonic acid, 4 minutes for *p*-aminophenylarsonic acid, and 7.5 minutes for inorganic arsenates. Inorganic arsenites are coeluted with dimethylarsenic acid; this mixture can be resolved with 5 mM $Na_4B_2O_7 \cdot 10H_2O$.

Quantification of the arsonic/arsenic acids is by reference for each compound to calibration curves, peak area, or height versus micrograms of arsenic. At concentrations corresponding to a 300 L sample containing 5–20 μg arsenic/m^3, recoveries exceeded 99%. When compared to neutron activation analysis and X-ray fluorescence spectrometry, arsenic recoveries from the arsonic/arsenic acids ranged from 90 to 120%.

5.3.2 Determination of Arsenic; NIOSH Method 7900

NIOSH method 7900 is applicable to the determination of arsenic in airborne particulates but not to volatile arsenic compounds such as As_2O_3 and AsH_3. The samples are collected on membrane filters and dissolved in a mixture of nitric, sulfuric, and perchloric acids; the arsenic contents are then quantified by atomic absorption spectrometry of arsine.

Airborne particulates from 500 L samples are collected at flow rates of 2 L/min on 37 mm diameter, 0.8 μm pore size cellulose ester membrane filters contained in cassette holders. The field blanks and the cassettes containing the filters with the samples are plugged and refrigerated during transport and storage.

The cassettes containing the samples and the field blanks are opened, and the membrane filters are transferred to 50 mL beakers. Beakers for spiked membrane filters and for reagent blanks are prepared. To each beaker is added 5 mL of 3:1:1 mixture of nitric, sulfuric, and perchloric acids. The beakers are covered with watch glasses, and the contents are heated on a 140°C hot plate until they become colorless. Dropwise addition nitric or perchloric acid may be necessary to complete dissolution of the samples. When the solutions have lost their color, the watch glasses are removed, and heating is continued until dense white fumes of SO_3 are evolved. The contents of the beakers are cooled, transferred to 25 mL volumetric flasks, and brought to volume with distilled/deionized water.

Arsenic is quantified by atomic absorption spectrometry of arsine at 193.7 nm in the hydrogen–argon flame by comparison to trivalent arsenic standards. Background correction with the deuterium continuum is necessary.

5.3.3 Determination of Arsenic Trioxide; NIOSH Method 7901

NIOSH Method 7901 is applicable to the determination of arsenic both in airborne particulate matter and in arsenic trioxide fumes. To assure retention of arsenic trioxide fumes, a specially prepared, sodium carbonate impregnated cellulose ester membrane filter and backup pad are used to collect the sample. The sample is dissolved with nitric acid and hydrogen peroxide, and the arsenic contents are quantified by electrothermal atomization–atomic absorption spectrometry.

The special preparation involves pipetting 250 μL of sodium carbonate–glycerol solution* on to 37 mm diameter, 0.8 μm pore size cellulose ester membrane filters with cellulose backup pads in cassette holders, drawing 30–60 L of clean air through the cassette, and allowing the filter to dry for 8 hours at 120°C or overnight at room temperature. The specially prepared filters should be used within one week.

Samples of up to 1000 L are collected at flow rates of 2 L/min. These are stable, but the determination of arsenic should not be delayed unnecessarily.

The filters and backup pads from the cassettes used to collect the samples and from field blanks are transferred to 50 mL beakers. The contents of each beaker are treated with 15 mL of concentrated nitric acid and covered with a watch glass. The reagent blanks and positive controls (spiked membrane filters) are similarly prepared. The contents of the beakers are heated on a 150°C hot plate until the volumes have been reduced to approximately 1 mL. After the contents have cooled and the watch glass bottoms and beaker sides have been rinsed into the residual acid, 1 mL of 30% hydrogen peroxide is added, and the solutions are heated to dryness on a steam bath. The dried residues are treated with exactly 10 mL of 1000 ppm Ni^{2+} solution and sonicated for 30 minutes. The arsenic contents of the resulting solutions are determined relative to arsenic standards by electrothermal atomization–atomic absorption spectrometry at a wavelength of 193.7 nm. The recommended dry–ash–atomize sequence is 70 seconds at 100°C, 30 seconds at 1300°C, and 10 seconds at 2700°C. Background corrections with the deuterium continuum are necessary.

5.3.4 Determination of Arsine; NIOSH Method 6001

In NIOSH method 6001 arsine and other volatile arsenic compounds are collected in tubes of activated charcoal preceded by membrane filters to remove airborne particulate matter. The arsine and other volatile arsenic compounds are recovered from the charcoal with nitric acid and quantified by electrothermal atomization–atomic absorption spectrometry.

Air samples of 10 L or less are collected at 0.1 L/min in commercially available, two-section (100 mg followed by 50 mg) activated charcoal tubes fitted with cellulose ester membrane prefilters. The samples are stable for at least 6 days when the tubes are closed with plastic caps.

To be sure that the capacity of the collection tube has not been exceeded, the front and back sections of the charcoal are placed in separate centrifuge tubes, treated with 10 mL of 0.01 M nitric acid, and sonicated for 30 minutes. Field blanks, charcoal from collection tubes spiked with known amounts of arsine, and reagent blanks are treated similarly. The nitric acid and charcoal phases are separated by centrifugation, and 50 μL aliquots of the liquid phases followed by 50 μL aliquots

*The sodium carbonate–glycerol solution is prepared by dissolving 9.5 g of Na_2CO_3 in 100 mL of distilled/deionized water and adding 5 mL of glycerin to the resulting solution.

of 1000 ppm Ni^{2+} solution are injected into the graphite furnace for quantification of arsenic by atomic absorption spectrometry. It can be assumed that the capacity of the collection tube has not been exceeded when the arsenic content of the rear section is no greater than one-tenth that of the front section. Recoveries of arsenic from 10 L air samples containing 0.094–0.404 mg/m^3 were greater than 90%, and breakthrough of arsenic to the rear section of the sampling tube did not occur after 4 hours of sampling air containing 0.405 mg/m^3 at a flow rate of 0.227 L/min.

5.3.5 Determination of Arsenic; US EPA Method 108

US EPA Method 108 is applicable to the determination of arsenic in particulate and gaseous airborne emissions from stationary sources.

Particulate and gaseous emissions are sampled isokinetically and collected on glass fiber filters and in liquid impingers. The arsenic is recovered, prepared for atomic absorption spectrometry, and quantified using flame, electrothermal, or gaseous hydride atomization.

The sampling train consists of a controlled-temperature filter holder containing a glass fiber filter followed by four Greenburg–Smith-type impingers. 100 mL of ASTM Type III water is added to each of the first three impingers, and 200 ± 0.5 g of indicating silica gel is added to the fourth impinger for determining the water content of the gaseous sample. Samples are collected at a flow rate below 28 L/min (1 cfm). The sample collection time should be sufficient to accurately determine maximum emissions during a 24-hour period. The temperature of the probe and filter holder should be maintained between 107 and 135°C during sample collection to prevent condensation.

At the end of the sampling period, the probe is removed from the duct or stack and allowed to cool. The cool probe is capped and, along with the impingers, is carried to a site reserved for transferring the samples to appropriate containers.

The glass fiber filter along with loose particulate material in the filter holder is transferred to a plastic sample container (container no. 1). The filter holder is rinsed with 0.1 M sodium hydroxide, and the washings are added to container no. 1.

The probe is washed with two 100 mL portions of 0.1 M sodium hydroxide, and the washings are added to a plastic sample container (container no. 2).

The silica gel in the fourth impinger is inspected to determine (by noting the color) whether its water capacity has been exceeded. It is then weighed to ascertain the water content of the gaseous sample. If the mass of the silica gel cannot be determined on site, the silica gel and any condensate present are transferred to a sample bottle (container no. 3) and transported to the laboratory for the determination of its mass.

The volumes of the solutions in the first three impingers are measured to ±1 mL. The solutions are transferred to a clean, one-liter, plastic sample container (container no. 4). All connecting tubing between the filter holder and the fourth impinger are rinsed with 0.1 M sodium hydroxide, and the rinsings are added to the liquid from the impingers in container no. 4.

Specimens of the 0.1 M sodium hydroxide should be retained as a field blank (container no. 5). Unused glass fiber filters should be retained for this purpose also.

The containers are sealed, labeled, and secured for shipment to the laboratory. Liquid levels are marked for later determinations of whether or not leakage has occurred.

The filters and loose particulates from container no. 1 are transferred to a 150 mL beaker and treated with 50 mL of 0.1 M sodium hydroxide. The contents of the beaker are warmed on a hot plate for 15 minutes, treated with 10 mL of nitric acid, brought to a boil, allowed to simmer for 15 minutes, and filtered through glass fiber into a clean, 150 mL beaker. The filtrate is evaporated to dryness, and the residue is dissolved in 5 mL of 50% (v/v) nitric acid and transferred to a 50 mL volumetric flask. The contents of the flask are brought to volume and retained for blending with the probe wash.

The volume of liquid in container no. 2 is checked to confirm the absence of leakage. The contents are filtered through glass fiber into a 200 mL volumetric flask and combined with the contents of the volumetric flask containing the solution prepared from the filters and loose particulates. The contents of the flask are diluted to exactly 200 mL with water, and a 50 mL aliquot is pipetted into a 150 mL beaker. The contents of the beaker are treated with 10 mL of nitric acid, brought to a boil, and evaporated to dryness. The residue is dissolved in 5 mL of 50% (v/v) nitric acid, and transferred to a 50 mL volumetric flask. The contents of the flask are brought to volume with water, and retained for atomic absorption spectrometry.

After the volume has been checked to confirm the absence of leakage, the liquid from the impingers and their washings in container no. 4 are transferred in 500 mL volumetric flask and brought to volume with water. A 50 mL aliquot is pipetted into a 150 mL beaker. The contents of the beaker are treated with 10 mL of nitric acid, brought to a boil, and evaporated to dryness. The residue is dissolved in 5 mL of 50% (v/v) nitric acid, transferred to a 50 mL volumetric flask, and brought to volume with water. The contents of the flask are retained for atomic absorption spectrometry.

A blank filter and the 0.1 M sodium hydroxide field blanks are prepared for atomic absorption spectrometry also.

Flame atomization–atomic absorption spectrometry may be used for quantification when the concentration is greater than 10 μg/mL. When the arsenic concentration is below 10 μg/mL, gaseous hydride or electrothermal atomization–atomic absorption spectrometry should be employed.

For the quantification of arsenic by atomic absorption in the air–acetylene or argon–hydrogen flame, measurements of samples and standards are made at 193 nm using an electrodeless discharge lamp as the source of resonance radiation. Background corrections are required for these measurements.

For quantification of arsenic by gaseous hydride atomization–atomic absorption spectrometry, an aliquot of the sample solution containing ~1–5 μg arsenic is transferred to the hydride generator reaction vessel and diluted with 15 mL of water. After additions of 15 mL of concentrated hydrochloric acid followed by 1 mL of 30% (m/v) potassium iodide solution have been made to the reaction vessel, it is heated in a 50°C water bath for 5 minutes, cooled, and connected to the hydride generator.

When the spectrometer, which has been fitted with an electrodeless discharge lamp for arsenic and a heated quartz cell for atomizing arsine, has stabilized at baseline, 5 mL of 5% sodium borohydride in 0.1 N sodium hydroxide solution is injected into the hydride generator, and the atomic absorbance of arsenic at 193 nm is recorded electronically. Quantification is achieved by direct comparison of background-corrected absorbances from the samples to those from arsenic standards.

Quantification of arsenic by electrothermal atomization–atomic absorption spectrometry requires the presence of nickel nitrate and hydrogen peroxide modifiers. To a 5 mL aliquot of sample solution, (arsenic content <1.5 μg) in a 10 mL volumetric flask are added 1 mL of 1% (m/v) nickel nitrate, 0.5 mL of 50% (v/v) nitric acid, and 1 mL of 3% hydrogen peroxide. The contents of the flask are brought to volume with water. Arsenic is quantified by comparison of background-corrected absorbances at 193 nm from standards and samples injected into the electrothermal atomizer. The dry–ash–atomize time–temperature sequence recommended by the manufacturer of the instrument should be followed, and an electrodeless discharge lamp should be used as a source of resonance radiation.

5.4 PROCEDURES FOR BARIUM; NIOSH METHOD 7056

Method 7056 is applicable to the collection, dissolution, and determination of water-soluble barium compounds in airborne particulate matter. The sample is collected on a cellulose ester membrane filter, dissolved in hot water, and its barium content is determined by flame atomic absorption spectrometry.

Airborne particulates from a 250 L sample are collected on a cellulose ester membrane filter having a diameter of 37 mm and a pore size of 0.8 μm at a flow rate of 2.5 L/min. The flow rate, temperature, and pressure should be measured and recorded at the beginning and at the end of the 100-minute sampling period.

In the laboratory, the filters are carefully removed from the cassettes, transferred to clean, labeled 50 mL beakers, and treated with 10 mL of boiling distilled water. The field blanks, filters spiked with known amounts of barium, and reagent blanks are treated similarly. The beakers are allowed to stand for 10 minutes, and their contents then are decanted into centrifuge tubes. The residues in the beakers are washed twice with 2 mL of hot distilled water, and the washings are added to the corresponding centrifuge tubes. The boiling water extraction/hot water washing cycle is repeated. Finally, the filters are removed with forceps and washed with streams of hot water so directed that the washings are added to the corresponding centrifuge tubes, and the original beakers are rinsed with 2 mL of hot water. The rinsings are added to the corresponding centrifuge tubes, which are centrifuged. The liquid phases are decanted into fresh beakers, treated with 3 drops of concentrated hydrochloric acid, and evaporated to volumes of approximately 0.5 mL. After cooling, 5 mL of 5% hydrochloric acid containing 1100 ppm Na^+ is added to each. When dissolution is assured, the contents of each beaker are transferred to 10 mL volumetric flasks and brought to volume with 5% hydrochloric acid containing 1100 ppm Na^+. The barium contents of the solutions are quantified relative to barium standards in 5% hydrochloric acid containing 1100 ppm Na^+ by atomic absorption spectrometry at 553.6 nm. Absorption measurements are made in the nitrous oxide–acetylene flame.

Collection efficiency from atmospheres containing aerosolized barium chloride solutions (0.3–1.1 mg/m^3) sampled at 1.4 L/min was 100%. Recovery of barium chloride from spiked filters was 102 ± 1.4%.

5.5 PROCEDURES FOR BERYLLIUM

In experiments with laboratory animals, the inhalation of beryllium and its compounds has produced cancers.[161] Consequently, both NIOSH and the US EPA have developed specifications for monitoring airborne beryllium concentrations.

5.5.1 Determination of Beryllium; NIOSH Method 7102

The collection and dissolution of airborne particulate matter for the determination of beryllium in workplace atmospheres is described in NIOSH Method 7102. The particulate matter is collected on a cellulose ester membrane filter and dissolved in a mixture of nitric and sulfuric acids. Its beryllium content is determined by electrothermal atomization–atomic absorption spectrometry.

Samples ranging in sizes from several tens of liters to several hundreds of liters are collected on cellulose ester membrane filters having diameters of 37 mm and pore size of 0.8 μm at flow rates of 1–4 L/min. The samples are stable after collection. Together with field blanks, the cassettes containing the sample filters are forwarded to the laboratory.

In the laboratory, the cassettes are opened carefully, and the field blank filters and those with the particulate matter are transferred to clean 125 mL beakers. Beakers with spiked filters and for the reagent blanks are prepared also. To the beakers are added 10 mL of concentrated nitric acid and 1 mL of concentrated sulfuric acid. The beakers are covered with watch glasses, and their contents are heated on a 140°C hot plate until the evolution of brown fumes ceases. The temperature of the hot plate is increased to 400°C and heating is continued until dense white fumes of SO_3 are evolved. After the contents of the beakers have cooled, the undersides of the watch glasses and the inside walls of the beakers are washed into the residual acid with distilled water. The contents of the beakers are evaporated to dryness without baking the residues. Exactly 10 mL of 2% m/v sodium sulfate 3% v/v sulfuric acid solution are added to each beaker. The contents of the beakers are heated in a 60–70°C water bath for 10 minutes and then allowed to stand at ambient temperature overnight to assure complete dissolution of beryllium sulfate.

Concentrations of beryllium in these solutions are determined by electrothermal atomization–atomic absorption spectrometry at 234.9 nm relative to beryllium standards in 2% Na_2SO_4/3% H_2SO_4 solution using a drying–ashing–atomization program of 20 seconds at 110°C, 10 seconds at 900°C, and 18 seconds at 2800°C with deuterium background correction. Recovery of beryllium from NIST SRM 2675 (Filter Media) exceeded 98%.

5.5.2 Determination of Beryllium; US EPA Method 103

Another US EPA method is applicable to the determination of beryllium in ducts and stacks at stationary sources. The samples are collected isokinetically on membrane filters.

Particulates are sampled isokinetically and collected on Millipore AA filter or equivalent filter demonstrating 99.95% efficiency for 0.3 μm dioctyl phthalate smoke. The sampling train consists of a stainless steel nozzle, a borosilicate glass lined probe, and a holder for the filter. The Millipore AA filter should be supported by a Whatman No. 41 filter to guard against breakage.

Ideally, samples should be collected at locations eight duct/stack diameters downstream and two duct/stack diameters up-stream from bends, expansions, constrictions, or other irregularities producing turbulence in the flow. Samples should be collected at distances 25, 50, and 75% across the duct or stack. The sample collection should be adequate to determine the maximum 24-hour beryllium emission. The minimum sampling rates is 14 L/min, and the minimum sampling time is 2 hours.

At the end of the sampling period, the sampling train should be removed to a clean sample recovery area. The Millipore filter and its Whatman No. 41 backing are carefully transferred to a clean sample container. Particulate matter in the probe liner and filter holder is washed into the sample container with acetone. The sample container is sealed and labeled. Field blanks for the filters and for the acetone are similarly prepared.

The requirement for beryllium quantification in US EPA Method 103 is flexible: "Any currently acceptable method such as atomic absorption, spectrographic, fluorometric, chromatographic or equivalent may be used." Hence, the preparation of the sample for analysis and the quantification of beryllium may be completed by the procedures described in Section 5.5.3.

5.5.3 Determination of Beryllium; US EPA Method 104

US EPA Method 104 is applicable to the determination of beryllium in duct and stack gases at stationary sources emitting gaseous effluents directly to the atmosphere without further processing. Samples of atmospheric emissions are collected isokinetically, and beryllium is retained on a membrane filter and in water-filled impingers. Beryllium is recovered from the filter and impingers and quantified by flame atomization atomic absorption spectrometry.

The sampling train consists of a borosilicate glass lined, temperature-controlled probe followed by a borosilicate glass, temperature-controlled filter holder containing a Millipore AA filter or equivalent capable of retaining 0.3 μm dioctyl phthalate smoke particles on a fritted glass support. The Millipore AA filter should be supported by a Whatman No. 41 filter to guard against breakage. The filter holder is followed by four Greenburg–Smith-type impingers. To the first and second impingers, separately, is added 100 mL of water; the third impinger is left empty, and 200 ± 0.5 g of indicating silica gel is added to the fourth impinger for determining the water content of the gaseous sample. Samples are collected at a flow rate below 28 L/minute (1 cfm) for a period of time sufficient "to accurately determine the maximum emissions that occur in a 24-hour period." To prevent condensation, the probe and the filter holder are maintained at 120 ± 14°C during sample collection.

At the end of the sampling period, the probe is removed from the duct or stack and allowed to cool. The cool probe is capped and, along with the impingers, is carried to a site reserved for transferring the samples to appropriate containers.

The Millipore AA filter and the Whatman No. 41 filter along with loose particulate matter in the filter holder are transferred to a sample bottle (container no. 1). The contents of the first three impingers are transferred to another sample bottle (container no. 2). All connecting tubing that goes to the fourth impinger is rinsed with water and acetone, and the washings are added to sample bottle containing the liquid from the impingers (container no. 2). Samples of the rinse water and rinse acetone should be retained as a field blank (container no. 3). The silica gel in the fourth impinger is inspected to determine (by noting the color) whether its water capacity has been exceeded. It is then weighed to within ± 0.5 g to find the water content of the gaseous sample. If the mass cannot be ascertained on site, the contents of the fourth impinger are transferred to a sample container (container no. 4) and retained for off-site measurement. The containers are sealed, labeled, and secured for shipment to the laboratory.

The filters and loose particulates from container no. 1 are transferred to a 150 mL beaker and treated with 35 mL of nitric acid. The contents of the beaker are heated on a hot plate until destruction of organic matter is complete. The remaining material is cooled, treated with 5 mL of sulfuric acid and 5 mL of perchloric acid, returned to the hot plate, and evaporated to dryness. The residue in the beaker is dissolved in exactly 10 mL of 25% (v/v) hydrochloric acid and retained for atomic absorption spectrometry.

The liquid from the impingers and their washings from container no. 2 are transferred in 100 mL increments to a 150 mL beaker and evaporated to dryness. When the last 100 mL increment has been evaporated to dryness, the residue is cooled and treated with 35 mL of nitric acid and heated until destruction of organic matter is complete. The contents of the beaker are treated with 5 mL of sulfuric acid and 5 mL of perchloric acid, returned to the hot plate, and evaporated to dryness. The residue in the beaker is dissolved in exactly 10 mL of 25% (v/v) hydrochloric acid and retained for atomic absorption spectrometry.

A blank filter and the water and acetone field blanks are prepared for atomic absorption spectrometry also.

Concentrations of beryllium in these solutions are determined by flame atomization–atomic absorption spectrometry at 234.9 nm in a rich nitrous oxide–acetylene flame relative to beryllium standards in 25% hydrochloric acid.

5.6 PROCEDURES FOR CADMIUM; NIOSH METHOD 7048

Method 7048 is applicable to the determination of cadmium in fumes and dusts. Samples are collected on cellulose ester membrane filters and dissolved by treatment with nitric and hydrochloric acids. Cadmium concentrations in the resulting solutions are measured by atomic absorption in the air–acetylene flame.

Dusts and fumes from samples of the workplace atmospheres ranging in size from 100 to 1000 liters are collected at flow rates of 1–3 L/min on cellulose ester membrane filters having a mean pore diameter of 0.8 μm. Although the samples on the filters in the cassettes are stable, they, along with the field blanks, should be returned to the laboratory without unnecessary delay.

In the laboratory, the filters are carefully transferred from the cassettes to 50 mL beakers, treated with 2 mL of concentrated nitric acid, and covered with watch glasses. Empty beakers similarly treated with nitric acid serve as reagent blanks. The contents of the beakers are heated on a 140°C hot plate until the volumes are reduced to about 0.5 mL. The cycle of addition of nitric acid/heating/volume reduction is repeated twice. After the third evaporation to roughly 0.5 mL, 2 mL of concentrated hydrochloric acid is added to the contents of each beaker and the beakers are heated as above. The cycle of addition of hydrochloric acid/heating/volume reduction is repeated twice. After the third evaporation of hydrochloric acid to about 0.5 mL, the contents of the beakers are cooled, treated with 10 mL of distilled water, and transferred to 25 mL volumetric flasks. The contents of the flasks are brought to volume with distilled water.

The cadmium concentrations of the resulting solutions are determined relative to standards in 0.5 M hydrochloric acid by atomic absorption spectrometry at 228.2 nm in the air–acetylene flame using the deuterium lamp for background correction. As an alternative to flame atomization, electrothermal atomization is permitted. The method was validated with 25 L samples using cadmium oxide dust (0.12–0.98 mg/m^3) and with 25 L samples of cadmium fume (0.04–0.18 mg/m^3).

5.7 PROCEDURES FOR CALCIUM; NIOSH METHOD 7020

In NIOSH Method 7020 for the determination of calcium in workplace fumes and dusts, samples are collected on cellulose ester membrane filters and dissolved by treatment with nitric and perchloric acids. Calcium concentrations in the resulting solutions are measured by atomic absorption in the air–acetylene flame.

Dusts and fumes from samples of the workplace atmospheres ranging in size from 50 to 250 L are collected at flow rates of 1–3 L/min on cellulose ester membrane filters having a mean pore diameter of 0.8 μm. Although the samples on the filters in the cassettes are stable, they, along with field blanks, should be submitted to the laboratory without unnecessary delay.

The cassettes are opened in the laboratory and the filters are carefully transferred to clean beakers. The contents of the beakers plus empty beakers serving as reagent blanks are treated with 5 mL of concentrated nitric acid, covered with watch glasses, and heated on a 140°C hot plate until most of the acid has been volatilized. Then to the contents of each beaker are added 2 mL of nitric acid and 1 mL of perchloric acid, and the beakers are heated on a 400°C hot plate until dense white fumes of chlorine heptoxide are evolved. The undersides of the watch glasses and the inside walls of the beakers are rinsed to the bottoms of the beakers, the beakers are returned to the 400°C hot plate, and their contents taken to dryness. The beakers are cooled, and the residues are dissolved in 5 mL of 5% hydrochloric acid, transferred to 100 mL (or smaller) volumetric flasks containing 2 mL of cesium nitrates ionization suppressor (73.40 g of $CsNO_3$/L) and 2 mL of lanthanum nitrate releaser [156 g La $(NO_3)_3 \cdot 6H_2O$/L], and brought to volume with 5% hydrochloric acid.

The calcium concentrations of the resulting solutions are determined relative to standards containing the ionization suppressor and releaser in 0.5 M hydrochloric acid by atomic absorption spectrometry at 422.7 nm in a lean air–acetylene flame.

The method was validated with 85 L samples using calcium oxide dust (2.6–10.2 mg/m^3). Recoveries were 100%; mean results showed a 0.4% negative bias.

5.8 PROCEDURES FOR CHROMIUM; NIOSH METHOD 7024

NIOSH Method 7024 is applicable to the determination of total chromium in workplace atmospheres. Samples are collected on cellulose ester membrane filters and dissolved by treatment with nitric and hydrochloric acids. Chromium concentrations in the resulting solutions are measured by atomic absorption in the nitrous oxide–acetylene flame.

Hundred- to thousand-liter samples are collected from workplace atmospheres at flow rates ranging from 1 to 3 L/min on cellulose ester membrane filters having a mean pore diameter of 0.8 μm. The samples on the filters in the cassettes are stable, but they, along with field blanks, should be returned to the laboratory without unnecessary delay.

In the laboratory, the filters are carefully transferred from the cassettes to clean beakers, treated with 2 mL of hydrochloric acid, and covered with watch glasses. The contents of the beakers are heated on a 140°C hot plate until the volumes are reduced to about 0.5 mL. The cycle of addition of hydrochloric acid/heating/volume reduction is repeated twice. After the third evaporation to about 0.5 mL, 3 mL of nitric acid is added to the contents of each beaker, and the beakers are heated as before. The cycle of addition of nitric acid/heating/volume reduction is repeated twice. After the third evaporation of nitric acid to about 0.5 mL, the contents of the beakers are cooled, transferred to 15 mL centrifuge tubes, and brought to volume with distilled water.

The chromium concentrations of the resulting solutions are determined relative to standards in 0.5% nitric acid by atomic absorption spectrometry at 357.9 nm in the nitrous oxide–acetylene flame. A rich air–acetylene flame may be used as an alternative to the nitrous oxide–acetylene flame. NIOSH Method 7024 was validated with samples collected from atmospheres produced by the thermal decomposition of chromium hexacarbonyl. Collection efficiency was 100%, and recoveries averaged 98% for samples containing 45–190 μg chromium.

5.9 PROCEDURES FOR COBALT; NIOSH METHOD 7027

NIOSH Method 7027 is directed to the determination of cobalt in airborne particulates from the occupational environment. Samples are collected on membrane filters and dissolved in aqua regia. The cobalt concentrations of the resulting solutions are determined by flame atomic absorption spectrometry.

Particulates from a 200 L air sample are collected using the filter and procedure described in Section 5.1. Although the samples on the filters in the cassettes are stable, they, along with field blanks, should be returned to the laboratory without unnecessary delay. In the laboratory, the filters with the samples, the field blanks, and spiked filters are transferred to clean beakers. To each of these plus each of those for the reagent blanks is added 3 mL of aqua regia. The beakers are covered with watch glasses and allowed to stand for 30 minutes before being heated on a 140°C hot plate until about 0.5 mL of acid remains. Then 3 mL of nitric acid is added, and the contents of the beakers are heated on a 140°C hot plate until about 0.5 mL of acid remains. A second 3 mL increment of nitric acid is added, the beakers returned to the 140°C hot plate, and their contents are evaporated to 1 mL. The undersides of the watch glasses are rinsed into the beakers with 5% nitric acid, and any residues are dissolved by the addition of a few milliliters of 5% nitric acid. The contents of the beakers are transferred to 10 mL volumetric flasks and brought to volume with 5% nitric acid.

The cobalt concentrations of the solutions are determined by atomic absorption spectrometry with background correction in the air–acetylene flame at 240.7 nm, by direct comparison to cobalt standards in 5% nitric acid.

Recoveries from filters spiked with 12–96 μg of cobalt were 98%.

5.10 PROCEDURES FOR COPPER; NIOSH METHOD 7029

NIOSH method 7029 is applicable to the determination of total copper from dust and fume in the workplace atmospheres, or, with slight modification, to the determination of soluble copper from dust in the presence of copper fume.

5.10.1 Determination of Total Copper (Dust and Fume)

Samples are collected on membrane filters and dissolved in nitric acid. Atomic absorption spectrometry is used for the quantification of total copper.

Fifty to 1500 L samples are collected on 0.8 μm cellulose ester filters at flow rates ranging from 1 to 3 L/min. The samples are stable, but unnecessary delay in initiating the analysis should be avoided.

In the laboratory, blank filters, spiked filters, and the filters containing the samples are transferred to clean beakers and treated with 6 mL of nitric acid. The beakers are covered with watch glasses and heated on a 140°C hot plate until the volumes of the contents have been reduced to about 0.5 mL. The contents of the beakers are treated with 2 mL of nitric acid, returned to the hot plate, and again heated until volume reduction to about 0.5 mL has been achieved. The cycle of addition of nitric acid/heating/volume reduction is repeated. After the third evaporation to about 0.5 mL, 2 mL of hydrochloric acid is added to the contents of each beaker, and the beakers are heated as just described. The cycle of addition of hydrochloric acid/heating/volume reduction is repeated twice without evaporation to dryness. After the third evaporation of hydrochloric acid to about 0.5 mL, the contents of the beakers are cooled, diluted with 10 mL of water, transferred to 25 mL volumetric flasks, and brought to volume with water.

The solutions in the volumetric flasks are aspirated into a lean air–acetylene flame for measurement of atomic absorption at 324.7 nm, and their copper concentrations are quantified by direct comparison to copper standards in 0.5 M hydrochloric acid.

5.10.2 Determination of Soluble Copper (Dust)

Samples of copper dust are collected as in Section 5.10.1

The separation of dust from fume begins with wetting a new, 47 mm diameter, 0.3 μm pore size cellulose membrane filter and placing it on a vacuum filtration apparatus. A blank filter, a spiked filter, or a filter containing a sample is placed, sample side up, on top of the filter on the filtration apparatus. Suction is applied and removed to remove air bubbles between the filters and to fully wet the upper filter. Another new, 47 mm diameter, 0.3 μm pore size cellulose membrane filter is placed over the filter containing the sample, then wetted and freed of air bubbles. Five milliliters of water is added to the filtration apparatus and drawn through the filters by application of vacuum. A second 5 mL of water s added and drawn through the filters. The filtrates are combined in a beaker and treated with 6 mL of nitric acid. The beakers are covered with watch glasses and heated on a 140°C hot plate until the volumes of the contents are reduced to about 0.5 mL. The residues in the beakers are taken through the cycles of nitric acid and hydrochloric acid heating/evaporation used in Section 5.10.1, and copper is quantified by the atomic absorption spectrometric procedure described in Section 5.10.1.

5.10.3 Determination of Copper Fume

It is possible to separately quantify soluble copper from dust and insoluble copper from fume in the same sample. After completion of the procedure described in Section 5.10.2, the three-filter "sandwich" is transferred to a clean beaker and treated with 6 mL of nitric acid. The beaker is covered with a watch glass and heated on a 140°C hot plate until the volume of the contents has been reduced to about 0.5 mL. The residue in the beaker is taken through the cycles of nitric acid and hydrochloric acid heating/evaporation used in Section 5.10.1, and copper is quantified by the atomic absorption spectrometric procedure described in Section 5.10.1.

The dust–fume separation was evaluated by first collecting fumes from a copper welding operation and then collecting dust from a copper sulfate mist generator. Recoveries of fume and dust were 96.5 and 94.7%, respectively.

5.11 PROCEDURES FOR LEAD

Both NIOSH and US EPA have published methodologies for determining airborne lead by atomic spectrometry.

5.11.1 Determination of Lead; NIOSH Method 7082

NIOSH Method 7082 is applicable to the determination of lead in the workplace atmosphere. Samples are collected on membrane filters, dissolved in a mixture of nitric acid and hydrogen peroxide, and analyzed for lead by flame atomization–atomic absorption spectrometry.

From 200 to 1500 L of air is filtered through a 0.8 μm cellulose ester membrane at a flow rate of 1–4 L/min. The samples, along with blank filters and spiked filters, are transported to the laboratory in capped cassettes. The materials are stable.

The filters are transferred to clean beakers and treated with 3 mL of nitric acid and 1 mL of hydrogen peroxide. In the absence of lead dioxide, hydrogen peroxide may be omitted. The beakers are covered with watch glasses, and the contents are heated on a 140°C hot plate until they have been reduced to about 0.5 mL. The cycle of adding nitric acid and hydrogen peroxide followed by heating and evaporation is repeated two times. The undersides of the watch glasses and the inside walls of the beakers are washed to the bottom with a few milliliters of 10% nitric acid, and the contents of the beakers are evaporated to dryness. The residues in the beakers are dissolved in 1 mL of nitric acid, transferred to 10 mL volumetric flasks, and brought to volume with water.

Sample preparation by microwave-assisted digestion with 5 mL of nitric acid at 180°C and 100 psi is allowed also.

Aspirations from the solutions in the volumetric flasks and from standards in 10% nitric acid are made into a lean air–acetylene flame, and the background-corrected absorbances are measured at either 217.0 or 283.3 nm. Quantification of the lead concentrations in the sample solutions is by direct comparison.

Collection efficiency of 0.8 μm cellulose ester membrane filters was 100%, and lead recovery from filters containing lead, lead oxide, and lead dioxide was better than 90%.

5.11.2 Determination of Lead; NIOSH Method 7105

With NIOSH Method 7105, samples are collected and prepared by the procedures described in Section 5.11.1. Quantification, however, is by electrothermal atomization–atomic absorption spectrometry. Twenty μL of sample or standard and 10 μL of matrix modifier [0.2 g $NH_4H_2PO_4$, 0.2 g Mg $(NO_3)_2$, and 1 mL HNO_3/100 mL] are injected into a pyrolytic graphite tube, and absorbance measurements are made with background correction at either 217.0 or 283.3 nm. A dry/ash/atomize cycle of 70 seconds at 110°C, 30 seconds at 800°C, and 5 seconds at 1500°C is recommended.

5.11.3 Determination of Lead; US EPA Reference Method[162]

Lead in particulate matter collected from ambient air on glass fiber filters is solubilized either by extraction with hot nitric acid or by sonication with nitric acid–hydrochloric acid mixture and quantified by atomic absorption spectrometry in the air–acetylene flame.

Particulates (mean aerodynamic diameters, 0.3 μm to 25–50 μm) are collected on preweighed, 8 in. × 10 in. glass fiber filters using Hi-Vol samplers operated at 1.5 m^3/min for 1440 minutes (24 hours). Clean filters are carefully installed in the samplers, and the samplers are run for at least 5 minutes; then measurements of flow rates and ambient temperatures and pressures are made and recorded. At the end of the 24-hour sampling period, flow rates and ambient temperatures and pressures are again measured and recorded. The samplers are stopped, and the loaded filters are carefully removed from the samplers, folded in half so that the loaded surfaces are on the insides, and placed in clean envelopes for transport to the laboratory along with the field blanks.

After equilibration at 25°C and 1 atm in the laboratory, the filters are reweighed to determine the masses of particulates collected. An 0.75 in. × 8 in. strip is then cut from the exposed area of each filter, using a template and a circular (pizza) cutter. The strips are folded in half twice and placed in 150 mL beakers. Lead is solubilized from the strips by either extraction with hot nitric acid or sonication with nitric acid–hydrochloric acid mixture.

For the hot nitric acid extraction, 15 mL of 3 M nitric acid is added to each beaker, the beakers are covered with watch glasses, and the contents are heated on a hot plate for 30 minutes. The nitric acid extracts are cooled to room temperature and transferred to 100 mL volumetric flasks. The bottoms of the watch glasses and the inside walls of the beakers are rinsed with distilled water, and the rinsings are transferred to the corresponding volumetric flasks. To each beaker is added 40 mL of distilled water. The beakers are covered with watch glasses and allowed to stand for at least 30 minutes to assure diffusion of residual solubilized lead from the glass fiber to the distilled water. These additions are transferred to the corresponding volumetric flasks. The contents of the flasks are mixed thoroughly and brought to volume with distilled water.

For the sonication, 15 mL of a solution 2.6 M in nitric acid and 0.9 M in hydrochloric acid is added to each beaker, and the beakers are covered with Parafilm or the equivalent. The contents of the beakers are sonicated for 30 minutes. The solutions are then transferred to 100 mL volumetric flasks. The Parafilm covers and the inside walls of the beakers are rinsed with distilled water, and the rinsings are transferred to the corresponding volumetric flasks. To each beaker is added 20 mL of distilled water. The beakers are covered with Parafilm and allowed to stand for at least 30 minutes to assure diffusion of residual solubilized lead from the glass fiber to the distilled water. These additions are transferred to the corresponding volumetric flasks. The contents of the flasks are mixed thoroughly and brought to volume with distilled water.

The lead concentrations of the solutions are determined by atomic absorption spectrometry in the air–acetylene flame at either 217 or 283 nm relative to standards in corresponding acid matrices. Background corrections, particularly at the shorter wavelength, may be necessary to compensate for differences between sample and standard solutions in total dissolved solids content. The MIBK-APDC extraction is an alternate to background correction. Intra- and interlaboratory precisions were 5 and 10%, respectively.

5.12 PROCEDURES FOR MERCURY

Mercury is encountered in workplace atmospheres and in the gaseous exhaust from various stationary sources. Methods of monitoring for both cases are described in Sections 5.12.1–5.12.4.

5.12.1 Determination of Mercury; NIOSH Method 6009

NIOSH Method 6009 is applicable to the determination of mercury in the workplace atmosphere. Mercury is collected in a sorbent tube, desorbed and dissolved with aqua regia, and quantified by cold vapor atomization–atomic absorption spectrometry. The method can be expanded to include separation and determination of particulate mercury and mercury vapor by inclusion in the sampling line of a cellulose ester prefilter.

A 10–100 L sample is collected on a Hopcalite sorbent tube at a flow rate of between 0.15 and 0.25 L/min. If separate determinations of particulate mercury and mercury vapor are to be made, a cassette containing an 0.8 μm cellulose ester membrane is installed upstream of the sorbent tube. The sample has been reported to be stable for 30 days at 25°C. The possibility of contamination or loss during storage can be avoided by beginning the determination of mercury promptly.

The sorbent tube, the front glass wool plug from the sampler, and the membrane filter if one is used are transferred to separate 50 mL volumetric flasks. Unused tubes and spiked tubes are placed in separate flasks also. To each flask are added 2.5 mL of nitric acid and 2.5 mL of hydrochloric acid. The contents of the flasks are allowed to stand at room temperature until the Hopcalite has dissolved. The brown solutions in the flasks become blue or blue-green when brought to volume with water.

A 20 mL aliquot is transferred from the volumetric flask to the reaction vessel of the cold vapor generator and diluted to 100 mL with water. In some generators, a 300 mL BOD bottle serves as the reaction vessel. Purge gas flow through the reaction vessel and spectrophotometer absorption cell is begun. The wavelength is set at 253.3 nm, the spectrophotometer is adjusted to zero, and 5 mL of 10% (m/v) tin(II) chloride solution is injected into the reaction vessel. The mercury vapor is swept from the reaction vessel into the absorption cell for measurement of its atomic absorption. The measurement is recorded electronically, and the mass of mercury in the sample is quantified by direct comparison to mercury standards treated in the same manner.

The analysis of 18 samples collected with Hopcalite or Hydrar sorbent tubes from atmospheres of mercury vapor showed a mean mercury recovery of 99%. No changes in mercury content were observed with sorbent tubes containing 0.3–9 μg during storage for 3 weeks at room temperature or during storage for 3 months at –15°C.

5.12.2 Determination of Mercury; US EPA Method 101

US EPA Method 101 is applicable to the determination of particulate and gaseous mercury in emissions from chloralkali production facilities and other installations where the duct and stack streams are primarily air. Samples of atmospheric emissions are collected isokinetically, and mercury is retained in impingers filled with iodine monochloride solution. Mercury is recovered from the impingers and quantified by cold vapor atomization atomic absorption spectrometry.

The sampling train consists of a quartz-lined, temperature-controlled probe followed by four Greenburg–Smith-type impingers. To begin, 100 mL of 0.1 M iodine monochloride solution* is added to each of the first three impingers; 200 ± 0.5 g of indicating silica gel is added to the fourth impinger, to allow determination of the water content of the gaseous sample. Samples are collected at a flow rate of 28 L/min (1 cfm) for 1440 minutes (24 hours).

At the end of the sampling period, the probe is removed from the duct or stack and allowed to cool. The cool probe is capped, and it, along with the impingers, is carried to a site reserved for transferring the samples to appropriate containers.

The volumes of the solutions in the first three impingers are measured to ± 1 mL and transferred to a clean, one-liter glass sample bottle with a ground glass neck and stopper. This is container no. 1. Materials in the probe are washed into container no. 1 with two 50 mL portions of 0.1 M ICl. Materials in the glass tubing connecting the components of the sampling train as far back as the inlet to the third impinger are washed into container no. 1 with 400 mL of ASTM Type II water. The bottle is stoppered and labeled, and the liquid level is marked to allow later determination of whether leakage has occurred.

*Procedure: 800 mL of hydrochloric acid is added to 800 mL of 25% (m/v) potassium iodide solution and cooled to room temperature. Then 135 g of potassium iodate is added, and the mixture is stirred until all the liberated iodine has dissolved. After dilution to 1800 mL with water, the solution is clear red-orange and in 1.0 M in ICl. It should be stored in a brown glass bottle. To obtain 0.1 M ICl, as is required for use in the impingers, 100 mL of the red-orange, 1.0 M solution is diluted to one liter with ASTM Type II water.

The silica gel in the fourth impinger is inspected to determine (by noting the color) whether its water capacity has been exceeded. It is then weighed to ascertain the water content of the gaseous sample. If the mass of the silica gel cannot be determined on site, the silica gel is transferred to a sample bottle (container no. 2) and transported to the laboratory for the determination of its mass.

A field blank of 50 mL 0.1 M ICl is prepared in a 100 mL glass sample bottle. This is container no. 3.

The samples are returned to the laboratory without delay. In the laboratory, the contents of container no. 1 are transferred to a 1-liter volumetric flask and brought to volume with ASTM Type II water. A 2 mL aliquot is transferred from the 1-liter flask to a 250 mL volumetric flask, treated with 10 mL of 5% (v/v) sulfuric acid, and brought to volume with Type II water. The diluted sample is stable for at least 72 hours.

Next 50 mL of Type II water is added to the reaction vessel of the cold vapor generator. A 5 mL aliquot of the diluted sample solution in the 250 mL volumetric flask is transferred to the reaction vessel of the cold vapor generator. A 300 mL BOD bottle may serve as the reaction vessel in some generators. Purge gas flow through the reaction vessel and spectrophotometer absorption cell is begun at 1.5 L/min. The spectrophotometer is adjusted to zero at a wavelength of 253.3 nm, and 5 mL of tin(II) chloride or tin(II) sulfate solution (20 g of the former; 25 g of the latter, dissolved in 25 mL hydrochloric acid and diluted to 250 mL) is injected into the reaction vessel. The mercury vapor is swept from the reaction vessel into the absorption cell for measurement of its atomic absorbance. The measurement is recorded electronically, and the mass of mercury in the sample is quantified by direct comparison to mercury standards treated in the same manner. The quality control program should include measurement of one blank and one standard after every five samples.

5.12.3 Determination of mercury; US EPA Method 101A

US EPA Method 101A is applicable to the determination of particulate and gaseous mercury emissions from sewage sludge incinerators and similar facilities. Samples of atmospheric emissions are collected isokinetically, and mercury is retained on a filter and in impingers containing potassium permanganate solution. Mercury is recovered from the impingers and quantified by cold vapor atomization–atomic absorption spectrometry.

The sampling train consists of a borosilicate glass lined, temperature-controlled probe followed by a borosilicate glass, temperature-controlled filter holder containing a glass fiber filter capable of retaining 0.3 μm dioctyl phthalate smoke particles on a stainless steel support followed by four Greenburg–Smith-type impingers. To the first, second, and third impingers are added 50, 100, and 100 mL, respectively, of potassium permanganate solution.* To permit determination of the water content

*It is customary to prepare 40 g potassium permanganate per liter of 10% (v/v) sulfuric acid fresh daily.

of the gaseous sample, 200 ± 0.5 g of indicating silica gel is added to the fourth impinger. Samples are collected at a flow rate of 28 L/min (1 cfm) for 1440 minutes (24 hours). The temperature of the filter holder is maintained at 120 ± 14°C during sample collection.

At the end of the sampling period, the probe is removed from the duct or stack and allowed to cool. The cool probe is capped, and it, along with the impingers, is carried to a site reserved for transferring the samples to appropriate containers.

The volumes of the solutions in the first three impingers are measured to ±1 mL and transferred to a clean, 1-liter glass sample bottle with a ground glass neck and stopper. This is container no. 1. Materials in the probe are washed into container no. 1 with 250–400 mL of the permanganate solution described earlier. Any brown residues of Mn_2O_3/MnO_2 present in the impingers are dissolved in a minimal volume of 8 N hydrochloric acid, and the resulting solution is added to the sample bottle. The sample bottle is stoppered and labeled, and the liquid level is marked to allow the bottle to be checked later for leakage.

The silica gel in the fourth impinger is inspected to determine (by noting the color) whether its water capacity has been exceeded. The unit is then weighed to determine the water content of the gaseous sample. If the mass of the silica gel cannot be determined on site, the silica gel is transferred to a sample bottle (container no. 2) and transported to the laboratory for this determination.

The glass fiber filter is carefully removed from the filter holder, folded in half so that the particulate deposit is on the inside, and transferred to a 100 mL glass sample bottle. This is container no. 3. The next step is to add 20–40 mL of the permanganate solution already described to container no. 3. The sample bottle is stoppered and labeled, and the liquid level is marked to allow the bottle to be checked later for leakage.

An unused glass fiber filter is folded in half and placed in a 100 mL glass sample bottle. This is container no. 4, to which 20–40 mL of the permanganate solution is added. The sample bottle is stoppered and labeled, and the liquid level is marked as before.

A field blank is prepared by adding 500 mL of the permanganate solution described earlier to a clean, 1-liter glass sample bottle with a ground glass neck and stopper. This is container no. 5. It is stoppered and labeled, and the liquid level is marked.

The samples are returned to the laboratory without delay. In the laboratory, the contents of container no. 1 are filtered through Whatman No. 40 paper into a volumetric flask. The filter is washed with 50 mL of the permanganate solution, and the washings are added to the volumetric flask. The contents of the flask are brought to volume with ASTM Type II water.

The contents of container no. 3 including the glass fiber filter are transferred to a 250 mL beaker and heated on a steam bath until most of the liquid has evaporated. Then 20 mL of nitric acid is added, the beaker is covered, and its contents heated on a 70°C hot plate for 2 hours. The contents of the beaker are filtered through Whatman No. 40 paper into a volumetric flask and brought to volume with water.

The contents of container no. 4, including the blank glass fiber filter, are transferred to a 250 mL beaker and treated by the same procedures described for the contents of container no. 3.

The contents of container no. 5 are filtered through Whatman No. 40 paper into a volumetric flask and treated by the same procedures described for the contents of container no. 1.

Next, 25 mL of type II water is added to the reaction vessel of the cold vapor generator. A 5 mL aliquot of the diluted sample solution in the 250 mL volumetric flask is transferred to the reaction vessel of the cold vapor generator. A 300 mL BOD bottle may serve as the reaction vessel in some generators. A mixture of 5 mL of 12% (m/v) sodium chloride and 12% (m/v) hydroxylamine chloride or hydroxylamine sulfate is added to the reaction vessel. Purge gas flow through the reaction vessel and spectrophotometer absorption cell is begun at 1.5 L/min. The spectrophotometer is adjusted to zero at a wavelength at 253.3 nm, and 5 mL of tin(II) chloride or tin(II) sulfate solution (20 g of the former; 25 g of the latter, dissolved in 25 mL hydrochloric acid and diluted to 250 mL) is injected into the reaction vessel. The mercury vapor is swept from the reaction vessel into the absorption cell for measurement of its atomic absorbance. The measurement is recorded electronically, and the mass of mercury in the sample is quantified by direct comparison to mercury standards pretreated with 5 mL of the permanganate solution described earlier and 5 mL of 15% (v/v) nitric acid solution to match the matrix of the sample. The quality control program should include measurement of one blank and one standard after every five samples.

5.12.4 Determination of Mercury; US EPA Method 102

US EPA Method 102 is applicable to the determination of particulate and gaseous mercury in the hydrogen streams of chloralkali production facilities. Samples of atmospheric emissions are collected isokinetically, and mercury is retained in impingers filled with iodine monochloride solution. Mercury is recovered from the impingers and quantified by cold vapor atomization–atomic absorption spectrometry.

Samples are collected by the procedures described in Section 5.12.2 with added emphasis on the hazards associated with the hydrogen atmospheres. Sample preparation and mercury quantification also are by the procedures described in Section 5.12.2.

5.13 PROCEDURES FOR NICKEL CARBONYL; NIOSH METHOD 6007

NIOSH Method 6007 is applicable to the determination of nickel carbonyl in the workplace atmosphere. Samples are collected on charcoal tubes, desorbed, dissolved with nitric acid, and the nickel carbonyl concentration of the atmosphere is quantified by electrothermal atomization–atomic absorption spectrometry of nickel. Potential interferences from other chemical forms of nickel are removed by protecting the charcoal tube with a prefilter.

The sample (8–80 L) is collected at a flow rate of 0.05–0.2 L/min using a two-section charcoal sorbent tube protected with a 0.8 μm cellulose ester membrane prefilter. Sample stability evaluations show 95% recovery after storage for 17 days at room temperature. Unnecessary delays should be avoided in transporting the samples and field blanks to the laboratory.

The front and back sections of the sorbent tube are placed in separate 2 mL vials and treated with 1 mL of 3% (v/v) nitric acid. Charcoal sections from spiked sorbent tubes are treated in the same manner. The vials are closed with plastic-lined screw caps and placed in an ultrasonic water bath for 30 minutes.

Nickel is quantified by electrothermal atomization–atomic absorption spectrometry. Background-corrected absorbance measurements are made at 232.0 nm on 20 μL injections from the vials or from nickel standards. A dry/ash/atomize cycle of 30 seconds at 110°C, 15 seconds at 800°C, and 10 seconds at 2700°C is recommended. The use of argon purge gas is recommended, as well.

Mean nickel recovery from two dozen samples of nickel carbonyl–carbon monoxide atmospheres was 93%. Mean desorption efficiency from charcoal spiked with nickel nitrate was 93% also. The use of the cellulose ester prefilter did not influence absorption of nickel carbonyl in the charcoal.

5.14 PROCEDURES FOR TUNGSTEN; NIOSH METHOD 7074

NIOSH Method 7074 is applicable to the determination of tungsten in the workplace atmosphere. Samples are collected on cellulose ester membrane filters. Soluble tungsten is recovered by extraction of the filter with water, and insoluble tungsten is recovered by subsequent dissolution of the residue with hydrofluoric acid–nitric acid mixture. After a matrix adjustment, tungsten is quantified by flame atomization–atomic absorption spectrometry.

A sample (200–1000 L) is collected on a 37 mm diameter, 0.8 μm pore size cellulose ester membrane filter at a flow rate of 1–4 L/min. The sample is stable for at least 2 weeks at 25°C.

The separation of soluble tungsten begins with wetting a new, 47 mm diameter, 0.3 μm pore size cellulose membrane filter and placing it on a vacuum filtration apparatus. The cassettes are opened in the laboratory, and a blank filter, a spiked filter, or a filter containing a sample of the analyte is placed, sample side up, on top of the filter on the filtration apparatus. Water (3 mL) is added to the filtration apparatus and allowed to stand in contact with the sample (or blank or spiked filter) for 3 minutes; then it is drawn through the filter by application of vacuum. A second 3 mL of water is added, allowed to stand, and drawn through the filter. The filtrates are combined in a 10 mL volumetric flask, treated with 1 mL of 20% (m/v) sodium sulfate solution, and brought to volume with water.

The filter with the residue remaining from the extraction with water is transferred to a PTFE beaker, treated with 5 mL of nitric acid and 5 mL of hydrofluoric acid, covered with a PTFE watch glass, and digested on a 150°C hot plate for 10 minutes. The watch glass is removed, and the contents of the beaker evaporated to a few milliliters and finally to dryness at a reduced temperature. (At this point, cobalt may be recovered from the residue in the filtrate obtained by extraction with 10 mL of 1% hydrochloric acid and subsequent filtration through a membrane filter.) The residue is treated with 5 mL of nitric acid and 5 mL of hydrofluoric acid, covered with a PTFE watch glass, and again digested on a 150°C hot plate for 10 minutes. The watch glass is removed, and the contents of the beaker evaporated to 1 mL and then allowed to go to dryness at 100°C. The residue in the beaker is dissolved by heating with 2.5 mL of 0.5 M sodium hydroxide and 2.5 mL of 20% (m/v) sodium sulfate solution, transferred to a 25 mL volumetric flask, allowed to cool, and brought to volume with water.

The concentrations of tungsten in the volumetric flasks are quantified by direct comparison to standards from absorbance measurements in a rich nitrous oxide–acetylene flame at 255.1 nm. Tungsten recoveries from filters spiked with soluble and insoluble compounds showed the samples to be stable for 2 weeks.

5.15 PROCEDURES FOR ZINC; NIOSH METHOD 7030

NIOSH Method 7030 is applicable to the determination of zinc and its compounds in the occupational environment. Samples are collected on membrane filters and dissolved in nitric acid. Zinc is quantified by flame atomization–atomic absorption spectrometry.

Air samples ranging in size from 4 to 400 L are collected at flow rates of 1–3 L/min, using 37 mm diameter, 0.8 μm pore size cellulose ester membrane filters. Samples are stable on the filters.

Cassettes are opened in the laboratory, and filters are transferred to clean beakers. Spiked filters and filters serving as field blanks are transferred to beakers also. Nitric acid (6 mL) is added to each beaker, the beakers are covered with watch glasses, and the contents are heated on a 140°C hot plate until clear, pale yellow solutions are obtained. Additional nitric acid may be needed to complete destruction of organic material. The underside of the watch glass and the inside walls are rinsed into the beaker with 1% (v/v) nitric acid, and the contents are heated until their volume is reduced to 0.5 mL. The residue in the beaker is treated with 10 mL of 10% nitric acid, heated for 5 minutes to ensure complete dissolution of the zinc, transferred to a 100 mL volumetric flask, and brought to volume with water.

The zinc concentrations of the solutions in the volumetric flasks are quantified by direct comparison to standards from background-corrected absorbance measurements at 213.9 nm in a lean air–acetylene flame.

5.16 DETERMINATION OF METALS IN AMBIENT AIR

Description of methods for the determination of airborne metals in addition to those given in Sections 5.2–5.15 can be found in the 1983 edition of *Regulatory Compliance Monitoring by Atomic Absorption Spectroscopy.*[163] The referenced edition contains descriptions of methods for the determination of iron, magnesium, manganese, molybdenum, rhodium, selenium, tellurium, thallium, tin, and yttrium in airborne particulate matter.

6

METHODS FOR COMPLIANCE WATER QUALITY MONITORING

The methods in this chapter are recommended or required for monitoring potable water supplies, as well as surface waters, groundwaters, and wastewaters. The United States Environmental Protection Agency (US EPA) has selected and developed methods to provide guidance for laboratories engaged in water quality monitoring. The methods have undergone interlaboratory evaluations, and they have been found acceptable. In certain cases, however, a particular sample cannot be successfully evaluated by these procedures. The US EPA's *Manual of Methods for Chemical Analysis of Water and Wastes*[50] and technical notes on mandatory and recommended methods modifications[164] should be consulted if additional detail on these procedures is needed. Other important references for the analysis of water and wastewater are *Standard Methods for the Examination of Water and Wastes Water*[52] and the *ASTM Annual Book of Standards*.[165]

Many of these procedures (hereafter designated SM and ASTM) have been adopted for monitoring potable water pursuant to the Safe Drinking Water Act (PL 93-523) and for monitoring ambient water pursuant to the first Clean Water Act (PL 95-217). In addition, it should be borne in mind that the methods for water quality compliance monitoring undergo continual growth and development as scientific and technological advances make possible procedural and instrumental improvements.

The collection, preservation, and preparation of water samples for trace elements determinations by atomic spectrometry were described in Chapters 2–4.

6.1 DETERMINATION OF METALS AND TRACE ELEMENTS IN WATERS AND WASTES BY INDUCTIVELY COUPLED PLASMA–ATOMIC EMISSION SPECTROMETRY; US EPA METHOD 200.7

US EPA Method 200.7[103] is applicable to the determination of aluminum, antimony, arsenic, barium, beryllium, boron, cadmium, calcium, chromium, cobalt, copper, iron, lead, lithium, magnesium, manganese, mercury, molybdenum, nickel, phosphorus, potassium, selenium, silica, silver, sodium, strontium, thallium, tin, vanadium, and zinc in groundwaters, surface waters, and drinking waters.

Sample preparation precedures for water and wastes were given in Section 3.1.2.2 under "Water and Wastewater Samples."

The spectrometer is calibrated according to the manufacturer's instructions using the appropriate blank and standard solutions, and emissions from the samples are measured at the wavelengths recommended in Table 1.5 or at other prescribed wavelengths. In the course of the determinations, emissions from the laboratory performance check (LPC) solution should be measured after every tenth sample. Recoveries from the LPC should be 100 ± 5% of the expected values. Similarly, emissions from the spectral interference check (SIC) solution should be measured to verify the validity of the interelement spectral interference correction process. The method of standard additions described in Section 1.1.5.2 may be required for the analysis of new or unusual samples.

In a single laboratory using Method 200.7, spike recoveries of at least 90% were achieved from tap water, well water, and pond water. Spike recoveries of at least 80% were achieved from an industrial effluent and from a municipal wastewater treatment works effluent.

6.2 DETERMINATION OF TRACE ELEMENTS IN WATERS AND WASTES BY INDUCTIVELY COUPLED PLASMA–MASS SPECTROMETRY; US EPA METHOD 200.8

US EPA Method 200.8[27] describes procedures for the determination of aluminum, antimony, arsenic, barium, beryllium, cadmium, chromium, cobalt, copper, lead, manganese, mercury, molybdenum, nickel, selenium, silver, thallium, thorium, uranium, vanadium, and zinc in groundwaters, surface waters, and drinking waters, as well as in wastewaters, sludges, and solid wastes.

Sample preparation procedures for water and wastes were given in Section 3.1.2.2 under "Water and Wastewater Samples."

The inductively coupled plasma mass spectrometer is tuned and calibrated, and its performance is verified with standards. Some typical operating conditions are listed in Table 6.1, and some estimated detection limits (EDLs) and method detection limits (MDLs) are presented in Table 6.2. Recoveries of 100 or 200 μg/L spikes from wastewater treatment works effluent exceeded 95% for all elements listed in Table 6.2 except thorium.

TABLE 6.1 Operating Conditions for Inductively Coupled Plasma–Mass Spectrometry

Condition	Instrument: VG PlasmaQuad Type I
Plasma forward power	1.35 kW
Coolant flow rate	13.5 L/min
Auxiliary flow rate	0.60 L/min
Nebulizer flow rate	0.70 L/min
Solution uptake rate	0.60 mL/min
Spray chamber temperature	15°C
Detector mode	Pulse counting
Replicate integration	3
Mass range	8–240 amu
Dwell time	320 μs
Multichannel analyzer channels	2048
Scan sweeps	85
Acquisition time	3 min/sample

TABLE 6.2 Estimated Detection Limits (EDL) for Inductively Coupled Plasma–Mass Spectrometry

Element	Analytical Mass	EDL (μg/L)[a]	MDL (μg/L)[b]
Aluminum	27	0.05	1.0
Antimony	121	0.08	0.4
Arsenic	75	0.9	1.4
Barium	137	0.5	0.8
Beryllium	9	0.1	0.3
Cadmium	111	0.1	0.5
Chromium	52	0.07	0.9
Cobalt	59	0.03	0.09
Copper	63	0.03	0.5
Lead	206, 207, 208	0.08	0.6
Manganese	55	0.1	0.1
Molybdenum	98	0.1	0.3
Nickel	60	0.2	0.5
Selenium	82	5	7.9
Silver	107	0.05	0.1
Thallium	205	0.09	0.3
Thorium	232	0.03	0.1
Uranium	238	0.02	0.1
Vanadium	51	0.02	2.5
Zinc	66	0.2	1.8

[a]The EDL (3*s*) were estimated from seven replicate integrations of the blank, 1% (v/v) nitric acid, followed by calibration of the instrument with three replicate integrations of a multielement standard.
[b]The MDL concentrations were computed for original matrix with allowance for sample dilution during preparation.

6.3 DETERMINATION OF TRACE ELEMENTS BY STABILIZED TEMPERATURE GRAPHITE FURNACE ATOMIC ABSORPTION; US EPA METHOD 200.9

US EPA Method 200.9[103] is applicable to the determination of aluminum, antimony, arsenic, beryllium, cadmium, chromium, cobalt, copper, iron, lead, manganese, nickel, selenium, silver, thallium, tin, and zinc in drinking water, groundwater, surface water, and wastewater. It is also applicable to total recoverable elements in sludges and sediments as well as in solid wastes and biological tissues.

Sample preparation procedures for water and wastewater were given in Section 3.1.2.2 under "Water and Wastewater Samples." Samples of solid wastes and biological tissues are prepared by the procedures presented in Section 3.1.2.2 under "Solid Wastes, Sludge, Sediment, and Soil Samples" and "Botanical and Biological Samples," respectively.

Some recommended measurement parameters are listed in Table 6.3. To initiate the measurement, a 20 μL aliquot of sample solution and a 20 μL aliquot of the palladium nitrate–magnesium nitrate matrix modifier are injected onto the platform of the pyrolytic graphite tube of the electrothermal atomizer. The dry/ash/atomize sequence shown in Table 6.3 is interrupted with a cooling phase prior to atomization.

TABLE 6.3 Recommended Operating Conditions for the Determination of Trace Elements by Electrothermal Atomization–Atomic Absorption Spectrometry[a]

			Cycle of Temperatures (°C)[b]		
Element	Wavelength (nm)	Slit (mm)	Dry	Ash	Atomize
Aluminum	309.3	0.7	200	1700	2600
Antimony	217.6	0.7	200	1100	2000
Arsenic	193.7	0.7	200	1300	2200
Beryllium	234.9	0.7	200	1200	2500
Cadmium	228.8	0.7	200	800	1600
Cobalt	242.5	0.2	200	1400	2500
Chromium	357.9	0.7	200	1650	2600
Copper	324.8	0.7	200	1300	2600
Iron	248.3	0.2	200	1400	2400
Lead	283.3	0.7	200	1250	2000
Manganese	279.5	0.2	200	1400	2200
Nickel	232.0	0.2	200	1400	2500
Selenium	196.0	2.0	200	1000	2000
Silver	328.1	0.7	200	1000	1800
Tin	286.6	0.7	200	1400	2300
Thallium	276.8	0.7	200	1000	1600
Zinc	213.9	0.7	200	700	1800

[a]A 15 μg palladium nitrate/10 μg magnesium nitrate matrix modifier was added for each measurement.
[b]During the drying and ashing steps, 5% hydrogen in argon purge gas flowed at 300 mL/m. Gas flow was interrupted during atomization. Volume of analyte injected was 20 μL.

Flow of the argon–hydrogen purge gas interrupted during atomization. Quantification of the elements listed in Table 6.3 can be by direct comparison when it can been demonstrated that matrix interferences do not require the method of standard additions.

Recoveries from a well water sample spiked at 2.5–25 μg/L with 14 of the 17 elements listed in Table 6.3 exceeded 90%. Iron, manganese, and zinc spikes were not added to the sample. Recoveries from similarly spiked samples of potable water and wastewater also exceeded 90% for most metals.

6.4 PROCEDURES FOR ALUMINUM

Aluminum is conveniently determined in a rich nitrous oxide–acetylene flame at a wavelength of 309.3 nm. Alternate wavelengths are 308.2, 396.2, and 394.4 nm. Ionization interferences are corrected for by the addition of potassium chloride. The thermal stability of most aluminum compounds allows ashing temperatures of up to 1400°C for electrothermal atomization. Few interferences are encountered, but the use of nitrogen purge gas should be avoided.

Aluminum is ubiquitous. Hence, special attention must be given to avoiding contamination when trace amounts of aluminum are being determined.

6.4.1 Determination of Aluminum; US EPA Method 202.1

US EPA Method 202.1 is applicable to the determination of aluminum in the concentration range of 5–50 mg/L in wastewater samples prepared by the methods described under "Water and Wastewater Samples" in Section 3.1.2.2. Samples and standards are treated with potassium chloride solution (95 g KCl/L) at a rate of 2 mL/100 mL. Treated samples and standards are aspirated into a rich nitrous oxide–acetylene flame, and the absorbances at 309.3 nm are measured. The concentrations of aluminum in the samples are determined by direct comparison with the standards.

Natural water samples spiked with six concentrates containing varying amounts of aluminum, cadmium, chromium, iron, manganese, lead, and zinc were analyzed in some three dozen laboratories. Mean results for aluminum were within ±10% of the theoretical values in the concentration range of 600–1200 μg/L. At aluminum concentrations below 50 μg/L, the mean results showed pronounced positive biases (200–500%), probably as a result of the contamination likelihood cited in Section 6.4.

6.4.2 Determination of Aluminum; US EPA Method 202.2

US EPA Method 202.2 describes the determination of aluminum in the concentration range of 20–200 μg/L by electrothermal atomization–atomic absorption spectrometry. Wastewater samples are prepared by the methods described in Section 3.1.2.2. Atomic absorption measurements are made on 20 μL injections at 309.3

nm in nonpyrolytic graphite tubes with continuous flow of argon purge gas, by means of a dry/ash/atomize sequence of 30 seconds at 125°C, 30 seconds at 1300°C, and 10 seconds at 2700°C. Corrections for background are required. For every sample matrix encountered, it is necessary to demonstrate that the method of standard additions is not necessary if the aluminum is to be quantified by direct comparison.

6.4.3 Other Procedures for Aluminum

Amendments[166] to the Clean Water Act allow the determination of aluminum by SM 3111D, SM 3113B, SM 3120B,* and by ASTM D4190-88 in addition to US EPA Methods 202.1, 202.2, 200.7, 200.8, and 200.9.

6.5 PROCEDURES FOR ANTIMONY

The determination of antimony is made at a wavelength of 217.6 nm in the air–acetylene flame. Because of its greater light output and longer life, the electrodeless discharge lamp (EDL) is preferred to the hollow cathode lamp (HCL). Possible spectral interferences from lead can be avoided by measuring the antimony absorbance at an alternate wavelength of 231.1 nm. The regulatory agencies have not endorsed the use of hydride generation or the argon–hydrogen flame for the determination of antimony.

6.5.1 Determination of Antimony; US EPA Method 204.1

US EPA Method 204.1 employs atomic absorption spectrometry in the air–acetylene flame for the determination of antimony at concentrations ranging from 1 to 40 mg/L. The nitric acid digestion described in Section 3.1.2.2 under "Water and Wastewater Samples" is employed for the preparation of wastewater samples. These and matrix-matched antimony standards are aspirated into a lean air–acetylene flame, and the absorbances at 217.6 nm are measured. Antimony concentrations in the samples are determined by direct comparison to the standards.

Precision of US EPA Method 204.1 was evaluated with spiked samples of wastewater. For samples spiked at 5 and 15 mg/L, the coefficients of variation were 8 and 1%, respectively, and the recoveries of antimony were 96 and 97%, respectively.

6.5.2 Determination of Antimony; US EPA Method 204.2

US EPA Method 204.2 is subject to withdrawal. It is directed to the determination of antimony by electrothermal atomization–atomic absorption spectrometry in the

*SM 3111, SM 3113, and SM 3120 designations employ flame atomization–atomic absorption spectrometry, electrothermal atomization–atomic absorption spectrometry, and inductively coupled plasma–atomic emission spectrometry, respectively.

concentration range of 20–300 μg/L. Wastewater samples are prepared by the nitric acid digestion described in Section 3.1.2.2 and dissolved in 5 mL of (1+1) hydrochloric acid. The final solutions of samples and standards should contain 2% (v/v) nitric acid. Background-corrected absorbances from 20 μL injections into nonpyrolytic graphite tubes are measured with uninterrupted flow of argon purge gas using the dry/ash/atomize sequence of 30 seconds at 125 C, 30 seconds at 800°C, and 10 seconds at 2700°C. Antimony concentrations of the samples are determined by standard addition. Direct comparison may be employed if the absence of matrix interferences can be demonstrated.

6.5.3 Other Procedures for Antimony

Amendments to the Clean Water Act allow the determination of antimony by SM 3111B, SM 3113B, and SM 3120B in addition to US EPA Methods 204.1, 204.2, 200.7, 200.8, and 200.9.

6.6 PROCEDURES FOR ARSENIC

Samples for the determination of arsenic are usually prepared by acid digestion in the presence of strong oxidizing agents. The absorbance measurements are made at 193.7 nm. The EDL is preferred to the HCL for the reasons cited in Section 6.5. Arsenic is frequently converted to the gaseous hydride. Atomization is achieved by thermal decomposition in the argon–hydrogen flame or in a heated quartz absorption cell.

6.6.1 Determination of Arsenic; US EPA Method 206.3

US EPA Method 206.3 describes the determination of arsenic in water and wastewater samples by atomic absorption spectrometry of the gaseous hydride. A 25 mL aliquot of the prepared sample is transferred to the reaction vessel of the hydride generator and treated with 1 mL of 20% (m/v) potassium iodide solution and 0.5 mL of tin(II) chloride solution (100 g of $SnCl_2$ dissolved in 100 mL of hydrochloric acid) to reduce pentavalent arsenic to the trivalent state. The reaction vessel is connected to the argon purge supply of the hydride generator, and 1.5 mL of zinc slurry is quickly injected. Trivalent arsenic is reduced to gaseous arsine and flushed from the reaction vessel to the argon–hydrogen flame of the atomic absorption spectrometer, where it is atomized. The atomic absorption of arsenic at 193.7 nm is recorded. Concentration of the arsenic in the sample can be quantified by direct comparison when the absence of matrix effects can be demonstrated. Otherwise, the quantification must be by standard additions.

Solutions containing 5, 10, and 20 μg/mL of arsenic from arsenilic acid were analyzed in a single laboratory by means of US EPA Method 206.5. The respective standard deviations were ±0.3, ±0.9, and ±1.1 μg/L, and the corresponding recoveries were 94, 93, and 85%.

6.6.2 Determination of Arsenic: US EPA Method 206.5

US EPA Method 206.5 is applicable to the determination of total arsenic in wastewater samples. To 50 mL of well-mixed sample are added 7 mL of (1+1) sulfuric acid and 5 mL of nitric acid. The acidified sample is heated cautiously until a clear or pale yellow solution is obtained and dense white fumes of SO_3 are evolved. Additional 5 mL increments of nitric acid may be added to prevent charring of the sample, as evidenced by a darkening of the solution. The colorless/pale yellow solution is treated with 25 mL of water and heated until dense white fumes of SO_3 are again evolved. The solution is cooled, transferred to a 50 mL volumetric flask, treated with 20 mL of hydrochloric acid, and brought to volume with water. Arsenic is determined in a 25 mL aliquot as described in Section 6.6.1.

6.6.3 Determination of Arsenic; US EPA Method 206.2

The determination of arsenic in water and wastewater samples by electrothermal atomization–atomic absorption spectrometry is described in US EPA Method 206.2. A 100 mL specimen of well-mixed sample is transferred to a 250 mL beaker and treated with 1 mL of nitric acid and 2 mL of 30% hydrogen peroxide. The contents of the beaker are heated cautiously until the volume is reduced to slightly less than 50 mL, cooled, and adjusted to a final volume of 50 mL with water. A 5 mL aliquot of this solution is pipetted into a 10 mL volumetric flask, treated with 1 mL of 1% nickel nitrate matrix modifier, and brought to volume with high purity water.

The atomic absorptions of replicate 20 μL injections from the contents of the volumetric flask and from standards prepared with nitric acid, hydrogen peroxide, and nickel nitrate to match the samples are measured at 193.7 nm in nonpyrolytic graphite tubes with continuous argon purge flow using a dry/ash/sequence of 30 seconds at 125°C, 30 seconds at 1100°C, and 10 seconds at 2700°C. Arsenic may be quantified by direct comparison when matrix interferences are absent. When the absence of matrix interferences cannot be demonstrated, standard addition must be used.

In a single laboratory using US EPA Method 206.2, the standard deviations for replicate analysis of tap water samples spiked at 20, 50, and 100 μg As/L were ±0.7, ±1.1, and ±1.6; corresponding arsenic recoveries were 105, 106, and 101%. A single laboratory using a wastewater sample containing 15 μg As/L spiked with 2, 10, and 25 μg/L addition arsenic reported recoveries of 85, 90, and 88%, respectively. The coefficients of variation from a single laboratory for the unspiked and spiked samples were 8.8, 8.2, 5.4, and 8.7%, respectively.

6.6.4 Other Procedures for Arsenic

Amendments to the Clean Water Act allow the determination of arsenic by SM 3114B, SM 3113B, SM 3120B, and SM 3500, as well as by ASTM D-2972-93B and ASTM D-2972-93C, in addition to US EPA Methods 206.3, 206.5, 206.2, and

200.7. Interruption of the purge gas flow during atomization is recommended in SM 3113B. For the ASTM procedures, the nickel nitrate matrix modifier may be added to samples and standards as described earlier, or it may be added directly to the electrothermal atomizer at a rate of 20 μg Ni per 20 μL injection.

6.7 PROCEDURES FOR BARIUM

Barium is conveniently determined in a rich nitrous oxide–acetylene flame at a wavelength of 553.6 nm. To compensate for the ionization interferences to which this determination is subject, sodium or potassium is added to both samples and standards.

6.7.1 Determination of Barium; US EPA Method 208.1

US EPA Method 208.1 is applicable to the determination of barium in water and wastewater samples at concentrations ranging from 1 to 20 mg/L. Samples are prepared by the methods described in Section 3.1.2.2. under "Water and Wastewater Samples." Samples and standards are treated with potassium chloride solution (95 g KCl/L) at a rate of 2 mL/100 mL. Treated samples and standards are aspirated into a rich nitrous oxide–acetylene flame, and the absorbances at 553.6 nm are measured. The concentrations of barium in the samples are determined by direct comparison with the standards.

In an interfacility comparison involving 13 laboratories, the coefficients of variation for samples containing 500, 1000, and 5000 μg/L were 8.6, 7.2, and 1.4%, respectively. In a single laboratory analyzing wastewater spiked at 0.40 and 2.0 mg Ba/L, the respective coefficients of variation were 11 and 6.5%; the corresponding barium recoveries were 94 and 113%.

6.7.2 Determination of Barium; US EPA Method 208.2

US EPA method 208.2 describes the determination of barium in the concentration range of 10–200 μg/L by electrothermal atomization–atomic absorption spectrometry. Wastewater samples are prepared by the methods described in Section 3.1.2.2. Atomic absorption measurements are made on 20 μL injections at 553.6 nm in nonpyrolytic graphite tubes with continuous flow of argon purge gas using a dry/ash/atomize sequence of 30 seconds at 125°C, 30 seconds at 1200°C, and 10 seconds at 2800°C. Corrections for background are recommended. The determination of barium may be made by direct comparison if the absence of matrix effects can be demonstrated; otherwise, standard addition is required.

Tap water samples spiked at 0.5 and 1.0 mg/L were analyzed in a single laboratory by US EPA Method 208.2 after (1+9) dilution, and precision corresponded to coefficients of variation of 0.5 and 0.2%, respectively. Recoveries of barium at these concentrations were 96 and 102%.

6.7.3 Other Procedures for Barium

Amendments to the Clean Water Act allow the determination of barium by SM 3111D, SM 3113B, and SM 3120B, as well as by ASTM D-3645-93B, in addition to US EPA Methods 208.1, 208.2, and 208.7. The use of pyrolytic graphite tubes is recommended in SM 3113B.

6.8 PROCEDURES FOR BERYLLIUM

The determination of beryllium is made at a wavelength of 234.9 nm in a rich nitrous oxide–acetylene flame.

6.8.1 Determination of Beryllium; US EPA Method 210.1

US EPA Method 210.1 describes the determination of beryllium in wastewater. Samples are prepared by methods described in Section 3.1.2.2. under "Water and Wastewater Samples," and standards and samples are aspirated into a rich nitrous oxide–acetylene flame. Atomic absorbances are measured at 234.9 nm, and the beryllium concentration is quantified by direct comparison.

Coefficients of variation of 10, 2, and 0.8% were obtained in a single laboratory using US EPA Method 210.1 for the determination of beryllium in wastewater samples spiked at 0.01, 0.05, and 0.25 mg/L. Beryllium recoveries were 100, 98, and 97%, respectively.

6.8.2 Determination of Beryllium; US EPA Method 210.2

US EPA Method 210.2 describes the determination of beryllium by electrothermal atomic absorption spectrometry in wastewater after sample preparation by methods presented in Section 3.1.2.2. Replicate 20 μL injections into the nonpyrolytic graphite tube of samples and standards prepared in 0.5% (v/v) nitric acid are atomized using a dry/ash/atomize sequence of 30 seconds at 125°C, 30 seconds at 1000°C, and 10 seconds at 2800°C. Atomic absorption is measured at 234.9 nm with uninterrupted argon purge gas flow.

6.8.3 Other Procedures for Beryllium

Amendments to the Clean Water Act allow determination of beryllium by SM 3113B, SM 3120B, and ASTM D-3645-93B in addition to US EPA Methods 210.1, 210.2, 200.7, 200.8, and 200.9.

6.9 PROCEDURES FOR BORON

The atomic absorption sensitivity for boron is poor: 40 μg/mL for 1% absorbance at 249.7 nm in a rich nitrous oxide–acetylene flame. Sensitivity is enhanced in nonaqueous solvents or in 50% (v/v) aqueous methanol. Boron is better determined by inductively coupled plasma–atomic emission spectrometry as described in US EPA Method 200.7 (see Section 6.1). Boron may also be determined by ASTM 3120B.

6.10 PROCEDURES FOR CADMIUM

Cadmium is conveniently determined in the air–acetylene flame at a wavelength of 228.8 nm. For high concentrations of cadmium, excessive dilutions can be avoided by making measurements of atomic absorption at the alternate wavelength of 326.1 nm. Electrothermal atomization may be accompanied by loss of cadmium due to premature volatilization. Cadmium is rendered more refractory by the addition of matrix modifiers to both samples and standards.

6.10.1 Determination of Cadmium; US EPA Method 213.1

US EPA Method 213.1 describes the determination of cadmium by flame atomization–atomic absorption spectrometry. Sample preparation is described in Section 3.1.2.2 under "Water and Wastewater Samples." Samples are standards are aspirated into the air–acetylene flame under identical conditions, and the atomic absorbances are measured at 228.8 nm. Quantification of cadmium is by direct comparison.

Six synthetic concentrates containing varying amounts of aluminum, cadmium, chromium, copper, iron, manganese, lead, and zinc were used to spike a natural water. Some five dozen laboratories analyzed these spiked water samples. For samples spiked in the range of 0.075 mg Cd/L, the mean value of the results showed a 5% negative bias. The bias was approximately 20% positive for samples spiked at 0.015 mg Cd/L.

6.10.2 Determination of Cadmium; US EPA Method 213.2

The determination of cadmium by electrothermal atomization–atomic absorption spectrometry is described in US EPA Method 213.2. Sample preparation is described in Section 3.1.2.2 under "Water and Wastewater Samples." To 100 mL of prepared sample or standard in 0.5% (v/v) nitric acid is added 2 mL of 40% (m/v) diammonium monohydrogen phosphate solution. Replicate 20 μL aliquots are injected into a nonpyrolytic graphite tube in the electrothermal atomizer, which is programmed for a dry/ash/atomize sequence of 30 seconds at 125°C, 30 seconds at 500°C, and 10 seconds at 1900°C. Absorbance measurements are made at 228.8 nm with uninterrupted flow of the argon purge gas.

To quantify cadmium by direct comparison, it is necessary to verify for every sample matrix evaluated that standard addition is not required. If such verification is not performed, or if any sample matrix fails the test, quantification of cadmium must be by standard addition.

A tap water sample spiked with cadmium to concentrations of 0.0025, 0.0050, and 0.0100 mg/L showed coefficients of variation of 4, 3, and 3%, respectively; the cadmium recoveries were 96, 99, and 98%, respectively.

6.10.3 Other Procedures for Cadmium

Cadmium may be determined by SM 3111B, SM 3113B, SM 3120B, and ASTM D-3557-90 in addition to US EPA Methods 213.1, 213.2, 200.7, 200.8, and 200.9.

6.11 PROCEDURES FOR CALCIUM

The atomic absorption of calcium is usually measured at a wavelength of 422.7 nm in either the air–acetylene flame or the nitrous oxide–acetylene flame. An alternate wavelength is 239.9 nm. Chemical interferences in the former require the use of a lanthanum chloride releasing agent, and ionization interferences in the latter call for the use of a potassium chloride suppressing agent.

6.11.1 Determination of Calcium; US EPA Method 215.1

US EPA Method 215.1 describes the determination of calcium by flame atomization–atomic absorption spectrometry. Samples are prepared by the methods described in Section 3.1.2.2 under "Water and Wastewater Samples." Lanthanum chloride releaser (29 g of La_2O_3 dissolved in 250 mL of hydrochloric acid and diluted to a final volume of 500 mL) is added to samples and standards in the ratio of 10 parts sample or standard to 1 part releaser. After addition of the releaser, the samples and standards are aspirated into a lean air–acetylene flame for measurement of atomic absorption at 422.7 nm. Calcium concentrations of the samples are quantified by direct comparison.

Coefficients of variation for the determination of calcium at concentrations of 9 and 36 mg/L were 3.3 and 1.5%, respectively. Calcium recoveries were 99% in both cases.

6.11.2 Other Procedures for Calcium

A 1% EDTA solution can be used instead of the lanthanum chloride releaser. Measurement of atomic absorption in a rich nitrous oxide–acetylene flame is another alternative that eliminates chemical interferences. Ionization interferences in the nitrous oxide–acetylene flame are controlled by the addition of potassium chloride solution (95 g KCl/L) to samples and standards at a rate of 2 mL/100 mL. Calcium may be determined by SM 3111B and SA 3120B and by ASTM D-511-92, in addition to US EPA Methods 215.1 and 200.7.

6.12 PROCEDURES FOR CHROMIUM

Chromium is most frequently determined at a wavelength of 357.9 nm. Neon-filled hollow cathode lamps should be used instead of argon-filled hollow cathode lamps to avoid interference from the 357.7 nm argon line. Some alternate wavelengths for the determination of chromium are 359.4, 360.5, 425.4, 427.5, and 429.0 nm. Both the air–acetylene flame and the nitrous oxide–acetylene flame have been used. Sensitivity and selectivity depend on the oxidant-to-fuel ratio in the flame and the region of the flame in which the measurements are made. Fewer difficulties are encountered with electrothermal atomization, but carbide formation and possible volatilization losses, as chromyl chloride must be considered. Rubio et al.[167] have published an excellent review on the determination of chromium in environmental and biological samples by atomic absorption spectrometry.

6.12.1 Determination of Chromium; US EPA Method 218.1

US EPA Method 218.1 is applicable to the determination of chromium in water and wastewater. The working range of the method is 0.5–10 mg/L. Samples are prepared by the procedures described in Section 3.1.2.2 under "Water and Wastewater Samples." Samples and standards are aspirated into a rich nitrous oxide–acetylene flame, and the absorbances are measured at 357.9 nm. The chromium concentrations of the samples are determined by direct comparison.

Six concentrates containing varying amounts of aluminum, cadmium, chromium, copper, iron, manganese, lead, and zinc were used to spike a natural water. The spiked samples were analyzed for chromium by some six dozen laboratories. The mean value for the sample spiked with 0.358 mg Cr/L showed a 30% coefficient of variation and a 5% negative bias, and that spiked at 0.085 mg Cr/L showed a 30% coefficient of variation and a 7% negative bias.

6.12.2 Determination of Chromium; US EPA Method 218.2

The determination of chromium in water and wastewater by electrothermal atomization–atomic absorption spectrometry is described in US EPA Method 218.2. Sample preparation is given under "Water and Wastewater Samples" in Section 3.1.2.2. The working range of this method is 5–100 μg/L. To each 100 mL of prepared sample of standard in 0.5% (v/v) nitric acid are added 1 mL of 30% hydrogen peroxide and 1 mL of calcium nitrate solution [11.8 g $Ca(NO_3)_2 \cdot 4H_2O$/100 mL]. The former is added to reduce chromium to the trivalent state, and the latter is added to minimize variations from the suppressive effects of calcium. Background-corrected measurements of absorbances from replicate 20 μL injections are made in nonpyrolytic graphite tubes with uninterrupted argon purge gas flow, using a dry/ash/atomize sequence of 30 seconds at 125°C, 30 seconds at 1000°C, and 10 seconds at 2700°C. Quantification of chromium is by standard addition. Direct comparison may be used for quantification if it can be shown that standard addition is unnecessary. A single laboratory using US EPA Method 218.2 for the analysis of tap water samples spiked at 19, 48, and 77 μg Cr/L reported coefficients of variation of 0.5, 0.2, and 1% and chromium recoveries of 97, 101, and 102%, respectively.

6.12.3 Other Procedures for Chromium

US EPA Method 218.3 describes a chelation–extraction method using APDC and MIBK for the determination of total chromium in water and wastewater at concentrations ranging from 1 to 25 μg/L. US EPA Method 218.4 describes a modification of Method 218.3 that allows the determination of hexavalent chromium. SM 3113B and SM 3120B may be used in addition to US EPA Methods 218.1, 218.2, 218.3, 218.4, 200.7, 200.8, and 200.9 for the quantification of chromium.

6.13 PROCEDURES FOR COBALT

Cobalt is conveniently determined in the air–acetylene flame at a wavelength of 240.7 nm. The nonabsorbing, parasitic radiation from cobalt lines at 240.8 and 240.9 nm is responsible for the nonlinearity of the absorbance–concentration relationship. Some alternate wavelengths are 242.5, 241.2, 252.1, and 243.6 nm, but these may suffer from spectral interferences when a multielement lamp containing nickel is used. The nitrous oxide–acetylene flame may be used with a modest loss in sensitivity. Aside from suppressive effects due to high concentrations of nitric acid, no significant difficulties are encountered when cobalt is determined by electrothermal atomization–atomic absorption spectrometry.

6.13.1 Determination of Cobalt; US EPA Method 219.1

US EPA Method 219.1 is applicable to the determination of cobalt in wastewater at concentrations ranging from 0.5 to 5 mg/L. Sample preparation methods are described under "Water and Wastewater Samples" in Section 3.1.2.2. For cobalt concentrations below 0.1 mg/L, the APDC-MIBK extraction procedure described in Section 3.2.3 may be used to pretreat prepared samples prior to the determination of cobalt.

Samples and standards are aspirated into the air–acetylene flame, and the absorbances are measured at 240.7 nm. If the chelation–extraction pretreatment is used, both samples and standards must undergo the procedure described in Section 3.2.3. Quantification of cobalt is by direct comparison.

Using a sample of wastewater spiked at cobalt concentrations of 0.2, 1,0, and 5.0 mg/L, precision ratings corresponding to coefficients of variation of 6.5, 1, and 1%, respectively, were demonstrated by a single laboratory. Cobalt recoveries at these concentrations were 98, 98, and 97%.

6.13.2 Determination of Cobalt; US EPA Method 219.2

US EPA Method 219.2 describes the determination of cobalt by electrothermal atomization–atomic absorption spectrometry. The working range of the method is 5–100 μg/L. Samples are prepared by the methods described under "Water and Wastewater Samples" in Section 3.1.2.2. Replicate 20 μL injections of samples and standards are atomized in the electrothermal atomizer using a dry/ash/atomize sequence of 30 seconds at 125°C, 30 seconds at 900°C, 10 seconds at 2700°C, and the absorbances at 240.7 nm are measured with background correction. Measurements are made with uninterrupted argon purge gas flow in nonpyrolytic graphite tubes. Quantification is by standard addition unless it can be verified that matrix interferences are absent.

6.13.3 Other Procedures for Cobalt

Amendments to the Clean Water Act allow the determination of cobalt by SM 3111B, SM 3113B, SM 3120B, and ASTM D-3553-88 as well as by US EPA Methods 219.1, 219.2, and 200.7.

6.14 PROCEDURES FOR COPPER

The determination of copper is conveniently made at a wavelength of 324.7 nm in a lean air–acetylene flame. Alternate wavelengths are 327.4, 216.5, and 22.6 nm. The nitrous oxide–acetylene flame may be used, but there is a threefold decrease in sensitivity. When copper is determined by electrothermal atomization, losses due to nitric acid suppression and premature volatilization are possible.

6.14.1 Determination of Copper; US EPA Method 220.1

US EPA Method 220.1 describes the determination of copper in water and wastewater at concentrations ranging from 0.2 to 5 mg/L by flame atomization–atomic absorption spectrometry. Samples are prepared by the methods described under "Water and Wastewater Samples" in Section 3.1.2.2. When the copper concentration is below 0.050 mg/L, the APDC-MIBK extraction procedure described in Section 3.2.3 may be used to pretreat prepared samples prior aspirating the solution containing the treated samples and those containing similarly treated copper standards. Atomic absorption measurements are made in a lean air–acetylene flame at 324.7 nm, and the copper concentration in the sample is quantified by direct comparison.

Some seven dozen laboratories analyzed six natural water samples spiked with varying amounts of aluminum, cadmium, chromium, copper, iron, manganese, lead, and zinc. For the sample spiked at 315 μg Cu/L, the coefficient of variation was 17% and the mean copper results showed a 1.5% negative bias. The coefficient of variation and bias for the sample spiked at 67 μg Cu/L were 34 and 4% positive, respectively.

6.14.2 Determination of Copper; US EPA Method 220.2

US EPA Method 220.2 describes the determination of copper in water and wastewater samples by electrothermal atomization–atomic absorption spectrometry. The working range of the method is 5–100 μg Cu/L. The absorbances of replicate 20 μL injections of samples prepared by the methods described under "Water and Wastewater Samples" in Section 3.1.2.2 and of copper standards are measured at 324.7 nm using a dry/ash/atomize sequence of 30 seconds at 125°C, 30 seconds at 900°C, 10 seconds at 2700° C. Background corrections are necessary, and quantification of copper must be by the method of standard additions unless the absence of matrix interferences can be demonstrated.

6.14.3 Other Procedures for Copper

SM 3111B, SM 3113B, and SM 3120B, as well as ASTM D-1688-90A and ASTM D-1688-90D, may be used in addition to US EPA Methods 220.1, 220.2, 200.7, 200.8, and 200.9 for the quantification of copper.

6.15 PROCEDURES FOR GOLD

Gold is conveniently determined in the air–acetylene flame at 242.8 nm. An alternate wavelength is 267.6 nm. Use of the nitrous oxide–acetylene flame results in a fourfold loss in sensitivity. When electrothermal atomization is employed, the volatility of gold limits the ashing temperature to 500°C. Consequently, matrix interferences are frequently encountered when gold is determined by electrothermal atomization–atomic absorption spectrometry, and background corrections and standard additions are often required.

6.15.1 Determination of Gold; US EPA Method 231.1

US EPA Method 231.1 describes a sample preparation procedure and a flame atomization–atomic absorption spectrometric procedure for gold. The working range is 0.5–20 mg Au/L.

A 100 mL specimen of the well-mixed sample is transferred to a 250 mL beaker, treated with 3 mL of nitric acid, placed on a steam bath, and evaporated to near dryness. The residue in the beaker is cooled, treated with 4 mL of hydrochloric acid and 2 mL of nitric acid, covered with a watch glass, and returned to the steam bath. After the contents of the beaker have been heated for 30 minutes, the watch glass is removed and heating is continued until the contents are near dryness. The residue in the beaker is treated with 1 mL of (1+1) nitric acid and filtered through glass fiber into a 100 mL volumetric flask. The sample and gold standards in 0.5% (v/v) are aspirated into a lean air–acetylene flame, and the absorbances are measured at 242.8 nm. Quantification is by direct comparison.

6.15.2 Determination of Gold; US EPA Method 231.2

The determination of gold by electrothermal atomization–atomic absorption spectrometry is described in US EPA Method 231.2. The working range of this method is 5–100 μg Au/mL. Samples are prepared by aqua regia digestion, as described in Section 6.15.1. The absorbances from 20 μL injections of prepared sample and from 20 μL injections of gold standard in 0.5% (v/v) nitric acid are measured with background correction at 242.8 nm using a dry/ash/atomize sequence of 30 seconds at 125°C, 30 seconds at 600°C, and 10 seconds at 2700°C. The measurements are made with continuous flow of argon purge gas in a nonpyrolytic graphite tube. Quantification is by standard addition unless the absence of matrix effects can be demonstrated.

6.15.3 Other Procedures for Gold

Amendments to the Clean Water Act allow the determination of gold by SM 3111B in addition to US EPA Methods 231.1 and 231.2.

6.16 PROCEDURES FOR IRIDIUM

The determination of iridium is usually carried out at 264.0 nm in a rich air–acetylene flame. Alternate wavelengths are 208.9, 266.5, 237.3, 285.0, 250.3, and 254.4 nm. Numerous interferences are encountered in the air–acetylene flame. There are fewer interferences in the nitrous oxide–acetylene flame, but the sensitivity is poorer than that achieved in the air–acetylene flame by a factor of 10. Few problems are encountered when iridium is determined by electrothermal atomization–atomic absorption spectrometry.

6.16.1 Determination of Iridium; US EPA Method 235.1

US EPA Method 235.1 is applicable to the determination of iridium in wastewater at concentrations ranging from 20 to 500 mg/L. Samples are prepared by the aqua regia digestion employed for gold (Section 6.15.1). Prepared samples and standards in 0.5% (v/v) nitric acid are aspirated into a rich air–acetylene flame, and the background-corrected absorbances at 264.0 nm are recorded. The iridium concentrations of the samples are quantified by direct comparison.

6.16.2 Determination of Iridium; US EPA Method 235.2

US EPA Method 235.2 describes the determination of iridium in wastewater by electrothermal atomization–atomic absorption spectrometry. The working range of the method is 0.1–1.5 mg/L. Samples are prepared by the aqua regia digestion employed for gold (Section 6.15.1). Replicate 20 μL injections of prepared samples and standards in 0.5% (v/v) nitric acid are injected into a pyrolytic graphite tube in the electrothermal atomizer. A dry/ash/atomize sequence of 30 seconds at 125°C, 30 seconds at 600°C, and 10 seconds at 2800°C with continuous flow of the argon purge gas is employed to measure the background-corrected absorbance at 264.0 nm. Background correction is recommended, as is quantification of the iridium concentration of the sample by standard addition. Direct comparison may be used for quantifying iridium if it can be demonstrated that standard addition is not necessary.

6.16.3 Other Procedures for Iridium

Amendments to the Clean Water Act allow the determination of iridium by SM 3111B in addition to US EPA Methods 235.1 and 235.2.

6.17 PROCEDURES FOR IRON

Iron is frequently determined at 248.3 nm in a lean air–acetylene flame. Alternate wavelengths are 248.8, 271.9, 252.7, and 372.0 nm. The use of a lean air–acetylene flame reduces interference from nitric acid and from nickel. Stoichiometry of the air–acetylene flame has little effect on sensitivity. Sensitivity is some threefold poorer in the nitrous oxide–acetylene flame, and a potassium chloride ionization suppressor is required. The use of electrothermal atomization for the determination of iron presents no special difficulties.

6.17.1 Determination of Iron: US EPA Method 236.1

US EPA Method 236.1 is applicable to the determination of iron in water and wastewater at concentrations ranging from 0.3 to 5 mg/L. Samples are prepared by the procedures described under "Water and Wastewater Samples" in Section 3.1.2.2. Prepared samples and standards in the same acid matrix are aspirated into a lean air–acetylene flame, and the absorbances are measured at 248.3 nm. Iron concentrations of the samples are quantified by direct comparison.

The chelation–extraction procedure described in Section 3.2.3 may be employed to determine low concentrations of iron.

Some 75 laboratories analyzed natural water samples spiked with varying amounts of aluminum, cadmium, chromium, copper, iron, manganese, lead, and zinc. For the sample spiked at 0.75 mg Fe/mL, the coefficient of variation was 23%, and the mean results showed a 1% negative bias. The coefficient of variation for the sample spiked at 0.40 mg Fe/mL was 38%, and the mean value was within ±0.5% of the theoretical value.

6.17.2 Determination of Iron; US EPA Method 236.2

The determination of iron in water and wastewater by electrothermal atomization–atomic absorption spectrometry is described in US EPA Method 236.2. The range of this method is 5–100 μg/L. Methods for sample preparation are described under "Water and Wastewater Samples" in Section 3.1.2.2. Absorbances from replicate 20 μL injections of prepared samples and standards in 0.5% nitric acid are measured at 248.3 nm. A dry/ash/atomize sequence of 30 seconds at 125°C, 30 seconds at 1000°C, and 10 seconds at 2700°C is recommended, with continuous flow of the argon purge gas, use of nonpyrolytic graphite tubes, and background correction. Direct comparison may be used for quantifying iron if it can be demonstrated that standard addition is not necessary.

6.17.3 Other Procedures for Iron

The determination of iron in water and wastewater may be performed by SM 3111D, SM 3113B, SM 3120B, and ASTM 1068-90, as well as by US EPA Methods 236.1, 236.2, 200.7, and 200.9.

6.18 PROCEDURES FOR LEAD

Lead determinations by flame atomization–atomic absorption spectrometry are made with lesser sensitivities than those achieved in the determinations of many other metals. The measurements are usually made in the air–acetylene flame at 283.3 nm. Threefold poorer sensitivity is obtained when the atomic absorption of lead is measured in the nitrous oxide–acetylene flame. Alternatively, the 217.0 nm resonance line may be used for the measurement of atomic absorption by lead. Both hollow cathode lamps and electrodeless discharge lamps are available as sources of resonance radiation. The spectrum from a neon-filled HCL is the less complicated. With an argon-filled HCL, interferences may be encountered when atomic absorption by lead is measured at 217.0 nm. Sensitivity is greatly enhanced when lead is determined by electrothermal atomization–atomic absorption spectrometry, but the volatility of lead limits the ashing temperature to 500°C. Electrothermal atomization is vulnerable to matrix interferences. Background corrections and standard additions are frequently employed to overcome these difficulties.

6.18.1 Determination of Lead; US EPA Method 239.1

US EPA Method 239.1 describes the determination of lead in water and wastewater. The working range of this method is 1–20 mg Pb/L. Samples are prepared by the procedures described under "Water and Wastewater Samples" in Section 3.1.2.2. Prepared samples and standards in 1% (v/v) nitric acid are aspirated into the air–acetylene flame, and the absorbances are measured at 238.3 nm. Quantification of lead in the samples is by direct comparison.

The MCL for lead in drinking water is 0.050 mg/L, well below the working range of US EPA Method 239.1. The chelation–extraction procedure described in Section 3.2.3 may be used to measure lead at this concentration. Prepared samples and standards are extracted, the extracts are evaporated, the residues are redissolved, and the resulting solutions are aspirated into the air–acetylene flame. The absorbances are measured at 283.3 nm, and lead is quantified by direct comparison.

In an interlaboratory comparison, six samples of natural water spiked with varying amounts of aluminum, cadmium, chromium, copper, iron, manganese, lead, and zinc were analyzed at some 75 different locations. For the sample spiked at 0.350 mg Pb/mL, the mean results showed a coefficient of variation of 34% and a negative bias of 2%. The corresponding values for the samples spiked at 0.090 mg Pb/mL were 48 and 1% positive, respectively.

6.18.2 Determination of Lead; US EPA Method 239.2

US EPA Method 239.2 describes the electrothermal atomization procedure for the determination of lead in water and wastewater at concentrations ranging from 5 to 100 μg/L. Samples are prepared by the methods described under "Water and Wastewater Samples" in Section 3.1.2.2. The absorbances from replicate 20 μL injections of prepared samples and standards in 0.5% (v/v) nitric acid containing 1 mL/100 mL lanthanum nitrate releaser are measured in nonpyrolytic graphite tubes at 238.3 nm with background correction. The measurements are made with uninterrupted flow of argon purge gas using a dry/ash/atomize sequence of 30 seconds at 125°C, 30 seconds at 500°C, and 10 seconds at 2700°C. Measurements made at 217.0 nm from an EDL show improved sensitivity.

Using tap water spiked at lead concentrations of 25, 50, and 100 μg/L, a single laboratory obtained results with coefficients of variation of 5, 3, and 4%, respectively. Recoveries of lead at these concentrations were 88, 92, and 95% respectively.

6.18.3 Other Procedures for Lead

Both SM 3113B and ASTM 3559-90D may be used for the determination of lead in water and wastewater in addition to US EPA Methods 239.1, 239.2, 200.8, and 200.9.

6.19 PROCEDURES FOR MAGNESIUM

The determination of magnesium is frequently made in the air–acetylene flame at a wavelength of 285.2 nm. Interferences for aluminum and silicon are overcome with a lanthanum chloride releaser. Atomization in the nitrous oxide–acetylene flame also eliminates these interferences, but the sensitivity is some three-fold poorer in this flame.

6.19.1 Determination of Magnesium; US EPA Method 242.1

US EPA Method 242.1 describes the determination of magnesium in wastewater. The working range of the method is 0.02–0.5 mg/L. Sample preparation procedures are described under "Water and Wastewater Samples" in Section 3.1.2.2. An aliquot of prepared sample or standard is treated with one-tenth its volume of lanthanum chloride releaser (29 g La_2O_3 dissolved in 250 mL hydrochloric acid and diluted to a final volume of 500 mL) and aspirated into the air–acetylene flame. Absorbances are measured at 285.2 nm, and the magnesium concentration is quantified by direct comparison.

In a single laboratory analyzing distilled water spiked at 2.1 and 8.2 mg Mg/L, the coefficients of variation for the mean results were 4 and 2%, respectively. Magnesium recoveries were 100% from both solutions.

6.19.2 Other Procedures for Magnesium

SM 3111D, SM 3120B, and ASTM D-511-92 may be used for the determination of magnesium in water and wastewater in addition to US EPA Methods 242.1 and 200.7.

6.20 PROCEDURES FOR MANGANESE

Manganese is frequently determined at a wavelength of 279.5 nm in a rich air–acetylene flame. Alternate wavelengths are 279.8 and 280.1 nm. Manganese may be determined also in the nitrous oxide–acetylene flame, although sensitivity is threefold poorer. The determination of manganese is conveniently made by electrothermal atomization–atomic absorption spectrometry also.

6.20.1 Determination of Manganese; US EPA Method 243.1

US EPA Method 243.1 is applicable to the determination of manganese in wastewater. The working range of the method is 0.1–3 mg/L. Lower manganese concentrations may be determined by the chelation–extraction procedure described in Section 3.2.3 after the wastewater samples have been prepared by the procedures described under "Water and Wastewater Samples" in Section 3.1.2.2. Prepared samples and standards matched in terms of acid matrix (and chelation–extraction if so treated) are aspirated into the air–acetylene flame. The absorbances are measured at 279.5 nm, and manganese is quantified by direct comparison.

Some six dozen laboratories analyzed natural water samples spiked with varying amounts of aluminum, cadmium, chromium, copper, iron, manganese, lead, and zinc. For the sample spiked at 0.450 mg Mn/mL, the mean showed a positive bias of 1.8% with a coefficient of variation of 19%. The corresponding values for samples spiked at 0.095 mg Mn/mL were 0 and 30%.

6.20.2 Determination of Manganese; US EPA Method 243.2

US EPA Method 243.2 describes the determination of manganese in wastewater by electrothermal atomization–atomic absorption spectrometry. The optimum concentration range for this method is 1–30 μg/L. Samples are prepared by the procedures described under "Water and Wastewater Samples" in Section 3.1.2.2, and the absorbances from replicate 20 μL injections of samples and standards are measured at

279.5 nm using a dry/ash/atomize sequence of 30 seconds at 125°C, 30 seconds at 1000°C, and 10 seconds at 2700°C. Continuous flow of the argon purge gas is recommended, along with the use of nonpyrolytic graphite tubes and background correction. Unless it can be shown to be unnecessary, standard addition is used to quantify the manganese concentrations of the samples.

6.20.3 Other Procedures for Manganese

SM 3111B, SM 3113B, and SM 3120B may be used in addition to US EPA Methods 243.1, 243.2, 200.7, 200.8, and 200.9 for the determination of manganese in water and wastewater.

6.21 PROCEDURES FOR MERCURY

The volatility of mercury allows a unique approach to its identification and quantification. Cold vapor atomic absorption spectrometry is applicable only to mercury. This approach is described under "Cold Vapor Generators" in Section 1.1.3.4. Mercury is conveniently determined by cold vapor atomic absorption spectrometry at a wavelength of 253.7 nm. The volatility of mercury demands special handling and preparation procedures to avoid losses. The same property presents special problems for the determination of mercury by electrothermal atomization–atomic absorption spectrometry.

6.21.1 Determination of Mercury; US EPA Method 245.1

US EPA Method 245.1 describes the determination of mercury in drinking water, surface water, saline water, and wastewater. The working range of the method is 0.5–10 μg/mL. Samples and standards are prepared as follows.

A 100 mL specimen of well-mixed sample, or an appropriate aliquot of mercury standard diluted to 100 mL or 100 mL of Type II ASTM water for a reagent blank, is transferred to the reaction vessel of the cold vapor generator. A 300 mL BOD bottle is frequently used for this purpose. To this vessel are added 5 mL of sulfuric acid, 2.5 mL of nitric acid, 15 mL of 5% (m/v) potassium permanganate solution, and 50 mL of Type II water. The contents of the reaction vessel are mixed well and allowed to stand for 15 minutes. If the purple color is discharged during this time, additional increments of permanganate solution are added to obtain a purple coloration that persists for 15 minutes. Next 8 mL of 5% (m/v) potassium peroxydisulfate solution is added to the reaction vessel, the vessel is capped and covered with aluminum foil, and the vessel is heated in a 95°C water bath for 2 hours.

Reaction vessels containing samples, standards, and blanks are treated individually from this point on. The contents of the reaction vessels are cooled, treated with 6 mL of 12% (m/v) sodium chloride–12% (m/v) hydroxylamine sulfate (or hydroxylamine chloride) solution. The reaction vessel is immediately connected to the cold vapor generator, and the headspace is purged for 30 seconds to remove chlorine and other interferants. After 5 mL of 10% (m/v) tin(II) chloride or tin(II) sulfate solution in 0.5 N sulfuric acid has been injected into the reaction vessel, the mercury vapor is swept into the absorption cell of the spectrometer, where absorbance at 253.7 nm is recorded electronically. Mercury concentrations in the samples are quantified by direct comparison.

A single laboratory spiked a composite sample of Ohio River water having a background mercury content of 0.35 μg/L with mercury corresponding to additions of 1.0, 3.0, and 4.0 μg Hg/L. Standard deviations for replicate analyses of the spiked and unspiked specimens were ±0.14, ±0.10, ±0.08, and ±0.16 μg/L, and the mercury recoveries from the spiked samples were 89, 87, and 87%.

The results from an interlaboratory study of US EPA Method 245.1 are presented in Table 6.4.

6.21.2 Determination of Mercury; US EPA Method 245.2

US EPA Method 245.2 describes an automated cold vapor method for the determination of mercury in water. This method may be also applicable to wastewater including domestic sewage. A schematic of the manifold for this method is shown in Figure 6.1.

The coefficients of variation for standards containing 0.5, 1.0, 2.0, 5.0, 10, and 20 μg Hg/L were, respectively, 8, 7, 5, 4, and 4%. Recoveries of 87–117% were reported by a single laboratory using surface water samples spiked at 10 μg Hg/L with 10 organic mercury compounds.

6.21.3 Other Procedures for Mercury

Amendments to the Clean Water Act allow the determination of mercury by SM 3112B and by ASTM D-3223-91 in addition to US EPA Methods 245.1, 245.2, and 200.8.

TABLE 6.4 Accuracy and Precision from an Interlaboratory Comparison of US EPA Method 245.1

Number of Participants	Mean (μg/L)	Coefficient of Variation	True Value (μg/L)	Bias (%)
76	0.34	88	0.21	+66
80	0.414	70	0.27	+53
82	0.674	80	0.51	+32
77	0.70	55	0.60	+18
82	3.41	43	3.4	+0.34
79	3.81	29	4.1	−7.1
79	8.77	42	8.8	−0.4
78	9.10	39	9.6	−5.2

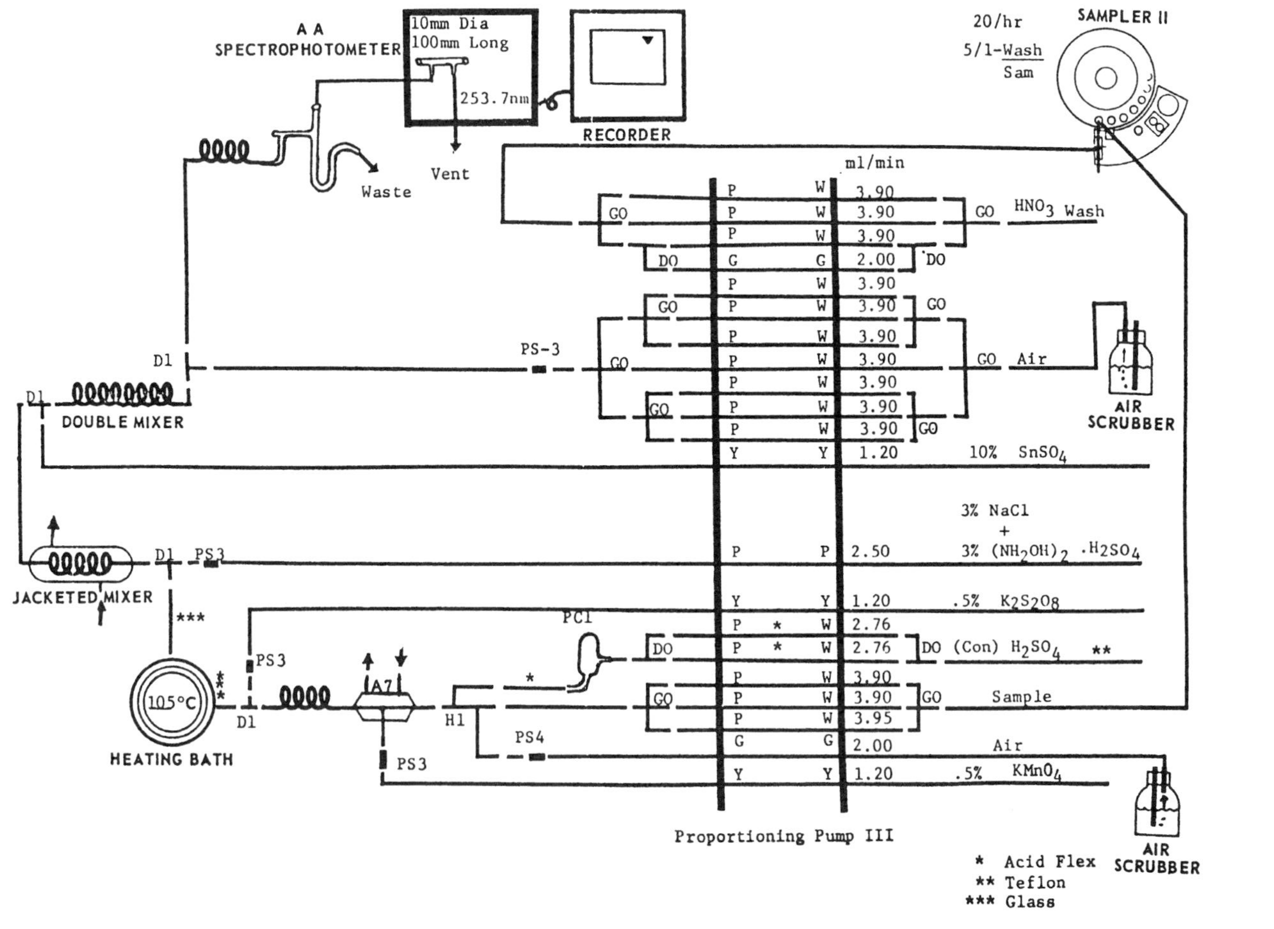

Figure 6.1. Mercury manifold schematic: P, purple; G, green; W, white, Y, yellow.

6.22 PROCEDURES FOR MOLYBDENUM

Molybdenum can be determined by flame atomization–atomic absorption spectrometry, but sensitivity is not as good as that obtainable with other metals. Absorbance measurements are made at 313.3 nm in rich nitrous oxide–acetylene flame. Alternate wavelengths are 317.0, 279.8, 319.4, 386.4, 390.3, and 315.8 nm. As with chromium, the sensitivity and selectivity with which molybdenum can be determined depend on the oxidant-to-fuel ratio in the flame and the region of the flame in which the measurements are made. There are numerous interferences. Some are mitigated by the addition of ammonium chloride or by chelation–extraction. While fewer interferences are encountered with electrothermal atomization, standard addition should be used for the quantification of molybdenum.

6.22.1 Determination of Molybdenum; ES EPA Method 246.1

US EPA Method 246.1 describes the determination of molybdenum in wastewater by flame atomization–atomic absorption spectrometry in the concentration range of 1–40 mg/L. Samples are prepared by the methods described under "Water and Wastewater Samples" in Section 3.1.2.2, and to each 100 mL of prepared sample or standard are added 2 mL of solution containing 139 g of $Al(NO_3)_3 \cdot 9H_2O$/200 mL. The treated samples and standards are aspirated into a rich nitrous oxide–acetylene flame, and the absorbances are measured at 313.3 nm. The molybdenum concentrations of the samples are quantified by direct comparison.

A single laboratory analyzed wastewater samples spiked to 0.30, 1.5, and 7.5 mg Mo/mL and obtained coefficients of variation of 2, 1, and 1%, respectively. Molybdenum recoveries were 100, 96, and 95%, respectively.

6.22.2 Determination of Molybdenum; US EPA Method 246.2

The determination of molybdenum by electrothermal atomization–atomic absorption spectrometry is described in US EPA Method 246.2. The working range of the method is 3–60 μg/L. The methods described in Section 3.1.2.2 under "Water and Wastewater Samples" are used to prepare the samples. The determination of molybdenum is by standard addition unless direct comparison can be justified. The samples and spiked samples should contain 0.5% (v/v) nitric acid. The absorbances of 20 μL injections are measured at 313.3 nm using a dry/ash/atomize sequence of 30 seconds at 125°C, 30 seconds at 1400°C, and 10 seconds at 2800°C.

6.22.3 Other Procedures for Molybdenum

SM 3111B, SM 3113B, and SM 3120B may be used for the determination of molybdenum in water and wastewater in addition to US EPA Methods 246.1, 246.2, and 200.7.

6.23 PROCEDURES FOR NICKEL

Nickel is most frequently determined in the air–acetylene flame at 232.0 nm using the narrowest spectral band pass possible. Parasitic radiations from the emission at 231.7 and 232.1 nm cause curvature of the absorbance–concentration relationship. Alternate wavelengths are 231.1, 352.5, 341.5, and 305.1 nm. Use of the nickel line at 352.5 nm is subject to interference from cobalt lines at 352.6 and 352.7 nm. Nickel sensitivity is some threefold poorer in the nitrous oxide–acetylene flame. The low volatility of most nickel compounds allows the use of high ashing temperatures in the electrothermal atomization sequence. While matrix interferences are reduced with electrothermal atomization, spectral interferences remain. Hence, standard addition is frequently required for the quantification of nickel.

6.23.1 Determination of Nickel; US EPA Method 249.1

Method 249.1 is applicable to the determination of nickel in wastewater over the concentration range of 0.3–5 mg/L. Samples are prepared in accord with the procedures described under "Water and Wastewater Samples" in Section 3.1.2. The chelation–extraction procedure described in Section 3.2.3 is recommended when the nickel concentrations are below 0.1 mg/L. The prepared and possibly extracted samples and standards are aspirated into the air–acetylene flame, and the absorbances are measured at 232.0 nm using a spectral bandwidth no greater than 0.2 nm. The calibration curve is prepared from the absorbances of the standards, and the nickel contents of the samples are determined by direct comparison.

In a single laboratory using mixed industrial and domestic effluents spiked at nickel levels of 0.2, 1, and 5 mg/L, the coefficients of variation were 6, 2, and 0.8%, respectively. The corresponding recoveries were 100, 97, and 93%.

6.23.2 Determination of Nickel; US EPA Method 249.2

Method 249.2 describes the determination of nickel in wastewater at concentrations ranging from 5 to 100 μg/L using electrothermal atomization. Samples are prepared in accord with the procedures of Section 3.1.2.2. To quantify nickel by direct comparison, it is necessary to verify for every sample matrix encountered that the method of standard addition is unnecessary. The absorbances of 20 μL injections of samples and spiked samples are measured at 232.0 nm using the following time–temperature program: dry, 30 seconds at 125°C; ash, 30 seconds at 900°C; atomize, 10 seconds at 2700°C.

6.23.3 Other Procedures for Nickel

Amendments to the Clean Water Act allow the determination of nickel by SM 3111B, SM 3113B, SM 3120B, and ASTM D-1886-90, in addition to US EPA Methods 249.1, 249.2, 200.7, 200.8, and 200.9.

6.24 PROCEDURES FOR OSMIUM

The sensitivity for the determination of osmium by atomic absorption spectrometry is poorer than that for most other metals. Absorbance measurements are made at 290.9 nm in a rich nitrous oxide–acetylene flame. Alternate wavelengths are 305.9, 263.7, 301.8, and 330.2 nm. Sensitivity is fivefold poorer in the air–acetylene flame.

6.24.1 Determination of Osmium; US EPA Method 252.1

Method 252.1 describes the determination of osmium in wastewater at concentrations ranging from 2 to 100 mg/L. Samples are prepared as follows.

A 100 mL aliquot of well-mixed sample is transferred to a 250 mL beaker and treated with 1 mL of nitric acid. The contents of the beaker are warmed on a hot plate for 15 minutes, cooled, and filtered into a 100 mL volumetric flask. Sulfuric acid (1 mL) is added, and the contents of the flask are brought to volume with distilled water. Samples and standards are aspirated into a rich nitrous oxide–acetylene flame, and the absorbances are measured at 290.9 nm. The concentrations of osmium in the samples are determined by direct comparison.

6.24.2 Determination of Osmium; US EPA Method 252.2

Method 252.2 describes the determination of osmium in wastewaters at concentrations ranging from 50 to 500 μg/L by electrothermal atomization–atomic absorption spectrometry. Samples are prepared by the procedure described in Section 6.24.1. For each sample matrix, the optimum ashing time and temperature must be established, and, to quantify osmium by direct comparison, the method of standard additions must be shown to be unnecessary. The absorbances of 20 μL injections of samples and spiked sample are measured at 290.9 nm. A 30-second drying stage at 105°C and a 10-second atomization stage at 2700°C are recommended. OSO_4 volatilizes at 150°C; hence, the ashing temperature must be determined on a matrix-by-matrix basis. Background correction, continuous argon flow, and pyrolytic graphite tubes are also recommended.

6.24.3 Other Procedures for Osmium

Amendments to the Clean Water Act do not allow alternative methods for the determination of osmium.

6.25 PROCEDURES FOR PALLADIUM

Palladium is frequently determined in a lean air–acetylene flame at 247.6 nm. Alternate wavelengths are 244.8, 276.3, and 240.5 nm. Palladium may also be determined in the nitrous oxide–acetylene flame, but there is a fivefold loss in sensitivity.

6.25.1 Determination of Palladium; US EPA Method 253.1

US EPA Method 253.1 describes the determination of palladium in wastewaters at concentrations ranging from 0.5 to 15 mg/L. The samples are prepared as follows.

A 100 mL aliquot of well-mixed sample is transferred to a 250 mL beaker, treated with 3 mL of nitric acid, and evaporated to near dryness on a steam bath. The contents of the beaker are cooled, treated with 3 mL of hydrochloric acid and 2 mL of nitric acid, covered with a watch glass, and returned to the steam bath for 30 minutes. The beaker is uncovered, and its contents evaporated to dryness. The residue on the walls of the beaker is washed to the bottom, and the contents of the beaker are filtered into a volumetric flask. The contents of the flask are treated with sufficient nitric acid to give a final concentration of 0.5% (v/v) and brought to volume with high purity water. Prepared sample and standards in 0.5% (v/v) nitric acid are aspirated into the air–acetylene flame, and the absorbances are measured at 247.6 nm. The quantification of palladium in the samples is by direct comparison.

6.25.2 Determination of Palladium; US EPA Method 253.2

US EPA Method 253.2 describes the determination of 20–400 μg/L palladium in wastewater by electrothermal atomization–atomic absorption spectrometry. Samples are prepared by the aqua regia digestion described in Section 6.25.1. The method of standard additions must be used unless it can be shown to be unnecessary, in which case, direct comparison may be used for quantification of palladium. The absorbances of replicate 20 μL injections of samples and spiked samples are measured at 247.6 nm using the following time–temperature program: dry, 30 seconds at 125°C; ash, 30 seconds at 1000°C; atomize; 10 seconds at 2800°C. Background correction, continuous purge argon, and pyrolytic graphite tubes are recommended.

6.25.3 Other Procedures for Palladium

Amendments to the Clean Water Act allow the determination of palladium in water and wastewater by SM 3111B in addition to US EPA Methods 253.1 and 253.2.

6.26 PROCEDURES FOR PHOSPHORUS

Phosphorus is among the elements determined by inductively coupled plasma–atomic emission spectrometry as described in US EPA Method 200.7.

6.27 PROCEDURES FOR PLATINUM

The determination of platinum is made at a wavelength of 265.9 nm in the air–acetylene flame. There are interferences by other noble metals, but these are overcome by the addition of lanthanum chloride or by means of the nitrous oxide–acetylene flame. Sensitivity is some fivefold poorer in this flame than in the air–acetylene flame. Alternate wavelengths are 306.5, 283.0, 293.0, 273.4, 270.2, 248.7, 299.8, and 271.9 nm.

6.27.1 Determination of Platinum; US EPA Method 255.1

US EPA Method 255.1 is applicable to the determination of platinum in wastewater at concentrations ranging from 5 to 75 mg/L. The samples are prepared with the aqua regia digestion described in Section 6.25.1. Prepared samples and standards are aspirated into the air–acetylene flame, and the absorbances are measured at 265.9 nm. Quantification of platinum is by direct comparison.

6.27.2 Determination of Platinum; US EPA Method 255.2

US EPA Method 255.2 describes the determination of platinum by electrothermal atomization–atomic absorption spectrometry in wastewaters at concentrations ranging from 0.1 to 2 mg/L. Samples are prepared by the aqua regia digestion described in Section 6.25.1. The method of standard additions must be used to quantify platinum unless direct comparison can be justified. The absorbances of 20 μL injections of samples and spiked samples or standards are measured at 265.9 nm using the following time–temperature program: dry, 30 seconds at 125°C; ash, 30 seconds at 1300°C; atomize, 10 seconds at 2800°C. Background correction, continuous flow of argon purge, and pyrolytic graphite tubes are recommended.

6.27.3 Other Procedures for Platinum

Amendments to the Clean Water Act allow the determination of platinum in water and wastewater by SM 3111B in addition to US EPA Methods 255.1 and 255.2.

6.28 PROCEDURES FOR POTASSIUM

Potassium is most frequently determined at 766.5 nm in the air–acetylene flame. An alternate wavelength is 769.9 nm. There are ionization interferences. Hence, both samples and standards are prepared for flame atomization–atomic absorption spectrometry in 0.1% sodium chloride. The photocathode material of some detectors is not highly responsive to the longer wavelengths. Therefore, some instruments may show poor sensitivity for potassium.

6.28.1 Determination of Potassium; US EPA Method 258.1

US EPA Method 258.1 is applicable to the determination of potassium in wastewater at concentrations ranging from 0.1 to 2 mg/L. The samples are prepared by the procedures described under "Water and Wastewater Samples" in Section

3.1.2.2. To each 100 mL of prepared sample and standard, 10 mL of 3% (m/v) sodium chloride is added, and absorbances are measured at 766.5 nm in the air–acetylene flame. The potassium concentrations of the samples are determined by direct comparision.

In a single laboratory using distilled water spiked at potassium levels of 1.6 and 6.3 mg/L, the coefficients of variation were 13 and 8%, respectively, and the corresponding potassium recoveries were 103 and 102%.

6.28.2 Other Procedures for Potassium

Amendments to the Clean Water Act allow the determination of potassium in water and wastewater by SM 3111B and SM 3120B in addition to US EPA Methods 258.1 and 200.7.

6.29 PROCEDURES FOR RHODIUM

Rhodium can be determined in the air–acetylene flame at 343.5 nm. Alternate wavelengths are 369.2, 339.7, 350.2, 365.8, and 370.1 nm. There are numerous interferences, many of which are eliminated in the nitrous oxide–acetylene flame. Use of this flame, however, results in a threefold loss of sensitivity relative to the air–acetylene flame.

6.29.1 Determination of Rhodium; US EPA Method 265.1

US EPA Method 265.1 is applicable to the determination of rhodium in wastewater at concentrations ranging from 1 to 30 mg/L. The samples are prepared by the aqua regia digestion described in Section 6.25.1. Prepared samples and standards are aspirated into the air–acetylene flame, and the absorbances at 343.5 nm are recorded. Quantification of rhodium is by direct comparison.

6.29.2 Determination of Rhodium; US EPA Method 265.2

US EPA Method 265.2 describes the determination by electrothermal atomization–atomic absorption spectrometry of rhodium in wastewater at concentrations ranging from 20 to 400 μg/L. Samples are prepared by the aqua regia digestion described in Section 6.25.1. If direct comparision for the quantification of rhodium is to be used, it must first be demonstrated that the method of standard additions is not needed. The absorbances of 20 μL injections of prepared samples and spiked samples or standards are measured at 343.5 nm using the following time–temperature program: dry, 30 seconds at 125°C; ash, 30 seconds at 1200°C; atomize, 10 seconds at 2800°C. Background correction is required for samples having high total dissolved solids. Continuous flow argon purge and pyrolytic graphite tubes are recommended.

6.29.3 Other Procedures for Rhodium

Amendments to the Clean Water Act allow the determination of rhodium in water and wastewater by SM 3111B in addition to US EPA Methods 265.1 and 265.2.

6.30 PROCEDURES FOR RUTHENIUM

Ruthenium can be determined in either the air–acetylene flame or in the nitrous oxide–acetylene flame at 349.9 nm. The former shows a sixfold better sensitivity than the latter, but more interferences. The addition of lanthium chloride eliminates the interferences and improves the sensitivity. Alternate wavelengths are 372.8 and 379.9 nm.

6.30.1 Determination of Ruthenium; US EPA Method 267.1

US EPA Method 267.1 describes sample preparation and flame atomization–atomic absorption spectrometry for the determination of ruthenium in wastewater at concentrations ranging from 1 to 50 mg/L. Samples are prepared as follows.

A 100 mL aliquot of well-mixed wastewater sample is transferred to a 250 mL beaker, treated with 2 mL of (1+1) hydrochloric acid, and warmed on a 95°C steam bath for 15 minutes. The contents of the beaker are cooled and filtered into a 100 mL volumetric flask. The contents of the flask are brought to volume with distilled water. The prepared sample and ruthenium standards are aspirated into the air–acetylene flame, and the absorbances are measured at 349.9 nm. Ruthenium is quantified by direct comparison.

6.30.2 Determination of Ruthenium; US EPA Method 267.2

US EPA Method 267.2 describes the determination of ruthenium by electrothermal atomization–atomic absorption spectrometry in wastewater at concentrations ranging from 0.1 to 2 mg/L. Samples are prepared by the procedure described in Section 6.30.1. To quantify ruthenium by direct comparison, it is necessary to demonstrate that the method of standard additions is not needed. The absorbances of replicate 20μL injections of prepared samples and spiked samples or standards are measured at 349.9 nm using the following time–temperature program: dry, 30 seconds at 125°C; ash, 30 seconds at 400°C; atomize, 10 seconds at 2800°C. Background corrections are required when the total dissolved solids are high. Continuous flow argon purge and pyrolytic graphite tubes are recommended.

6.30.3 Other Procedures for Ruthenium

Amendments to the Clean Water Act allow the determination of ruthenium in water and wastewater by SM 3111B in addition to US EPA Methods 267.1 and 267.2.

6.31 PROCEDURES FOR SELENIUM

The most commonly encountered procedures for the determination of selenium involve generation of the gaseous hydride, hydrogen selenide, and its subsequent thermal decomposition and atomization. Absorbance measurements are made at 196.0 nm. Alternate wavelengths are 204.0, 206.3, and 207.5 nm. Because of its transparency to short wavelength radiation, the argon–hydrogen flame is recommended for the decomposition–atomization. It is possible also to carry out the decomposition–atomization of the hydride in a quartz tube heated either by a resistance wire coil wrapping or by a flame. Because the flame used to heat the quartz tube it is not in the optical path, its composition is of little consequence to the absorbance measurements. The higher output intensity of the electrodeless discharge lamp makes it preferable to the hollow cathode lamp as a resonance radiation source. Although electrothermal atomization is 100 times more sensitive than gaseous hydride atomic absorption spectrometry, the volatility of selenium compounds limits ashing temperatures and, consequently, increases matrix interferences. The addition of a nickel nitrate matrix modifier, background correction, and standard addition are helpful in compensating for these interferences.

6.31.1 Determination of Selenium; US EPA Method 270.2

US EPA Method 270.2 describes electrothermal atomization–atomic absorption spectrometry for the determination of selenium in drinking water and in wastewater at concentrations ranging from 5 to 100 μg/L. The samples are prepared as follows.

A well-mixed, 100 mL aliquot of sample is transferred to a 250 mL beaker, treated with 1 mL of nitric acid and 2 mL of 30% hydrogen peroxide, and heated below the boiling point for 1 hour or until the volume is reduced by more than half. The contents of the beaker are cooled and diluted to 50 mL with distilled water.

A 5 mL aliquot is transferred from the beaker to a 10 mL volumetric flask, treated with 1 mL of 1% nickel nitrate matrix modifier [24.78 g $Ni(NO_3)_2 \cdot 6H_2O$/500 mL], and brought to volume with distilled water. The absorbances of replicate 20 μL injections from the volumetric flask and from selenium standards containing nitric acid, hydrogen peroxide, and nickel nitrate are measured at 196.0 nm using the following time–temperature program: dry, 30 seconds at 125°C; ash, 30 seconds at 1200°C; atomize, 10 seconds at 2700°C. The method of standard additions must be used for the quantification of selenium unless it can be demonstrated that direct comparison is interference free. Use of background corrections, continuous flow argon purge, and nonpyrolytic graphite tubes is recommended.

A recovery of 99% was obtained from an effluent spiked to a selenium concentration of 20 μg/L. For industrial effluents spiked to 50 μg/L, recoveries ranged from 94 to 112%. In a 0.1% nickel nitrate matrix with selenium concentrations of 5, 10, 20, 40, and 50 μg/L, the coefficients of variation for replicate measurements were 14.2, 11.6, 9.3, 7.2, 6.4, and 4.1%, respectively. In a single laboratory making replicate selenium determinations using tap water spiked at concentrations of 5, 10, and 20 μg/L, the coefficients of variation were 12, 4, and 2.5%, respectively. The corresponding recoveries were 92, 98, and 100%.

6.31.2 Determination of Selenium; US EPA Method 270.3

US EPA Method 270.3 describes the determination of selenium as the gaseous hydride. This method is applicable to the determination of selenium in drinking water, surface water, groundwater, saline water, and wastewater. The working range of Method 270.3 is 2–20 μg/L.

Samples are prepared by the same procedures described in Section 6.6.2 for the preparation of samples prior to the determination of arsenic by US EPA Method 206.5. The generation of hydrogen selenide and its subsequent atomization in the argon–hydrogen flame are essentially the same as those described in Section 6.6.1 for the determination of arsenic by Method 206.3. The absorbances of samples and standards at 196.0 nm are recorded electronically, and selenium in the samples is quantified by direct comparison. As a part of the data validation, samples of industrial wastewaters spiked with known amounts of selenium should be carried through the procedure to establish recovery factors.

A single laboratory determined selenium in replicate samples containing 5, 10, and 15 μg/L. The coefficients of variation were 12, 11, and 19%, respectively, and the corresponding recoveries were 100, 100, and 101%.

6.31.3 Other Procedures for Selenium

The determination of selenium may be made by SM 3113B and by ASTM D-3859-93D in addition to US EPA Methods 207.2, 207.3, 200.8, and 200.9.

6.32 PROCEDURES FOR SILICA

Silica is among the elements determined by inductively coupled plasma–atomic emission spectrometry as described in US EPA Method 200.7.

6.33 PROCEDURES FOR SILVER

Silver is conveniently determined in the air–acetylene flame at 328.1 nm. An alternate wavelength of 338.3 nm may be used with a 50% loss of sensitivity. Use of the nitrous oxide–acetylene flame results in a threefold loss in sensitivity. The volatility of silver compounds limits ashing temperature. Hence, electrothermal atomization–atomic absorption sepectrometry is subject to matrix interferences. The photosensitivity and limited solubility of silver chloride make necessary special procedures for sample collection, storage, and preparation.

6.33.1 Determination of Silver; US EPA Method 272.1

US EPA Method 272.1 is applicable to the determination of silver in drinking water and in wastewater at concentrations ranging from 0.1 to 4 mg/L. This method makes use of cyanogen iodide solution (4.0 mL ammonia, 6.5 g KCN, and 5.0 mL 1.0 N iodine in 50 mL of distilled water diluted to 100 mL) to recover silver compounds absorbed on the walls of the sample container and/or precipitated as silver chloride. The cyanogen iodide is used at a rate of 1 mL per 100 mL of sample *after* the sample has been rendered ammoniacal. The sample treated with cyanogen iodide is allowed to stand for 1 hour before proceeding with the determination. When samples are prepared by the procedures described under "Water and Wastewater Samples" in Section 3.1.2.2, the use of hydrochloric acid is avoided. For low levels of silver, the chelation–extraction procedure described in Section 3.2.3 may be employed. The prepared samples and standards (carried through the chelation–extraction procedure when it has been used) are aspirated into the air–acetylene flame, and the absorbances are measured at 328.1 nm. The silver contents of the samples are quantified by direct comparison.

A synthetic sample containing 50 μg/mL of silver was analyzed by 50 laboratories in a round robin. The coefficient of variation for the mean results was 17%, and the mean differed from the theoretical value by 11%.

6.33.2 Determination of Silver; US EPA Method 272.2

US EPA Method 272.2 describes the determination of silver at concentrations ranging from 1 to 25 μg/L in both drinking water and wastewater by electrothermal atomization–atomic absorption spectrometry. Absorbed and/or precipitated silver compounds are recovered with cyanogen iodide solution as described in Section 6.33.1, and samples are prepared by the procedures described in Section 3.1.2.2. The use of hydrochloric acid, however, is avoided. It is necessary to demonstrate that the method of standard additions is not needed before direct comparison may be used to quantify silver. The absorbances of 20 μL injections of prepared samples and spiked samples are measured at 328.1 nm using the following time–temperature program: 30 seconds drying at 125°C, 30 seconds ashing at 400°C, and 10 seconds atomizing at 2700°C. Background corrections, continuous flow of argon purge, and nonpyrolytic graphite tubes are recommended.

In a single laboratory using tap water spiked at silver levels of 25, 50, and 75 μg/L, the mean results showed the following coefficients of variation: 1.6, 1.4, and 1.2%. The silver recoveries from these samples were 94, 100, and 104%, respectively.

6.33.3 Other Procedures for Silver

SM 3111B, SM 3113B, and SM 3120B may be used for the determination of silver in addition to US EPA Methods 272.1, 272.2, 200.7, 200.8, and 200.9.

6.34 PROCEDURES FOR SODIUM

Sodium is conveniently determined in the air–acetylene flame at 589.0 nm. The effects of ionization interferences can be minimized by adding excess potassium chloride to both samples and standards. The effects of ionization interference can be decreased also by using cooler flames such as air–propane or air–hydrogen.

6.34.1 Determination of Sodium; US EPA Method 273.1

US EPA Method 273.1 is applicable to the determination of sodium in wastewater at concentrations ranging from 0.03 to 1 mg/L. Samples are prepared by the procedures described under "Water and Wastewater Samples" in Section 3.1.2.2. The prepared samples and the standards are treated with 2% (m/v) potassium chloride solution at a ratio of 100 mL of sample or standard to 1 mL of potassium chloride solution. The treated samples and standards are aspirated into the air–acetylene flame, and the absorbances are measured at 589.0 nm. The sodium concentrations in the samples are quantified by direct comparison.

In a single laboratory using distilled water spiked to sodium levels of 8.2 and 52 mg/L, the coefficients of variation for the mean results were 1.2 and 1.3%, respectively. The corresponding recoveries were 102 and 100%.

6.34.2 Other Procedures for Sodium

Sodium may be determined by SM 3111B in addition to US EPA Methods 273.1 and 200.7.

6.35 PROCEDURES FOR THALLIUM

Thallium is determined in the air–acetylene flame at a wavelength of 276.8 nm. Both the HCL and the EDL are available as sources of thallium resonance radiation. The latter gives higher light output and longer life than the former. Alternate wavelengths are 377.6 and 238.0 nm. Use of the nitrous oxide–acetylene flame results in fourfold lower sensitivity. The volatility of thallium compounds limits ashing temperatures to 500°C. Hence, electrothermal atomization–atomic absorption spectrometry of thallium suffers from matrix interferences, and standard addition is frequently required to compensate for them.

6.35.1 Determination of Thallium; US EPA Method 279.1

US EPA Method 279.1 is applicable to the determination of thallium in wastewater in the concentration range of 1–20 mg/L. Samples are prepared by the procedures described under "Water and Wastewater Samples" in Section 3.1.2.2, but the addition of hydrochloric acid is omitted. Samples and standards are aspirated into the air–acetylene flame, and the absorbances are measured at 276.8 nm. Thallium is quantified by direct comparison.

In a single laboratory using mixed industrial/domestic effluent spiked at thallium levels of 0.6, 3.0, and 15 mg/L, the coefficients of variation for the mean results were 3, 2, and 0.3%, respectively, and the corresponding thallium recoveries were 100, 98, and 98%.

6.35.2 Determination of Thallium; US EPA Method 279.2

US EPA Method 279.2 describes the determination of thallium in wastewater at concentrations ranging from 5 to 100 μg/L by electrothermal atomization–atomic absorption spectrometry. The samples are prepared by the procedures described in Section 3.1.2.2. The absorbances of replicate 20 μL injections of samples and spiked samples are measured at 276.8 nm. The following time–temperature program is recommended: dry, 30 seconds at 125°C; ash, 30 seconds at 400°C; atomize, 10 seconds at 2400°C. Background corrections, continuous flow argon purge gas, and nonpyrolytic graphite tubes are recommended also.

6.35.3 Other Procedures for Thallium

Thallium may be determined by SM 3111B and SM 3120B in addition to US EPA Methods 279.1, 279.2, and 200.7.

6.36 PROCEDURES FOR TIN

The determination of tin by flame atomic absorption spectrometry is made with significantly poorer sensitivity than that which is obtained for most other metals. Measurements are usually made in the nitrous oxide–acetylene flame at 286.3 nm. Alternate wavelengths are 224.6, 235.5, 270.6, 303.4, 254.7, 219.9, 300.9, and 233.5 nm. The air–acetylene flame gives results equal in sensitivity to the nitrous oxide–acetylene flame but with more interferences. Sensitivity is improved in cooler flames such as air–propane, but more interferences are encountered. Electrothermal atomization is capable of greater sensitivity, but the volatility of tin compounds limits the ashing temperature, which increases the potential for matrix interferences. Tin can be determined as the gaseous hydride with improved sensitivity and selectivity.

6.36.1 Determination of Tin; US EPA Method 282.1

Method 282.1 describes the determination of tin in wastewater at concentrations ranging from 10 to 300 mg/L. The samples are prepared by the procedures described under "Water and Wastewater Samples" in Section 3.1.2.2, and the prepared samples and standards are aspirated into the nitrous oxide–acetylene flame. Absorbance measurements are made at a wavelength of 286.3 nm, and tin is quantified by direct comparison.

In a single laboratory using mixed industrial/domestic effluent spiked at tin concentrations of 4, 20, and 60 mg/L, the coefficients of variation were 6, 3, and 1%. The corresponding recoveries of tin were 96, 101, and 101%.

6.36.2 Determination of Tin; US EPA Method 282.2

US EPA Method 282.2 describes the determination of tin in wastewater at concentrations ranging from 20 to 300 μg/L by electrothermal atomization–atomic absorption spectrometry. Sample preparation is by the procedures described in Section 3.1.2.2. If it can be shown that the method of standard additions is unnecessary, the determination of tin may be made by direct comparison. Replicate 20 μL aliquots of the prepared samples and prepared samples spiked with known amounts of tin are injected into the electrothermal atomizer, and the absorbances are measured at 224.6 nm. The time–temperature program is as follows: dry, 30 seconds at 125°C; ash, 30 seconds at 600°C; atomize, 10 seconds at 2700°C. Background corrections, continuous flow argon purge, and nonpyrolytic graphite tubes are recommended.

6.36.3 Other Procedures for Tin

Tin may be determined by SM 3111B and SM 3113B in addition to US EPA Methods 282.1, 282.2, and 200.7.

6.37 PROCEDURES FOR TITANIUM

The sensitivity with which titanium can be determined is poor in comparison to that obtainable for most other metals. Titanium is determined in a rich nitrous oxide–acetylene flame at 365.3 nm. Alternate wavelengths of almost equal sensitivity are 364.3, 320.2, 363.6, 335.5, 375.3, 334.2, 399.9, and 390.0 nm. Numerous elements interfere with the determination of titanium by enhancing the sensitivity. The addition of potassium chloride has been recommended to level these enhancement effects. The enhancement by fluoride is leveled by making standards and samples 0.1 M in ammonium fluoride.

6.37.1 Determination of Titanium; US EPA Method 283.1

US EPA Method 283.1 describes the sample preparation and flame atomization–atomic absorption spectrometry of titanium in wastewater at concentrations ranging from 5 to 100 mg/L. Sample preparation is as follows.

A 100 mL aliquot of well-mixed sample is transferred to a 250 mL beaker and treated with 2 mL of sulfuric acid and 3 mL of nitric acid. The beaker is covered with a watch glass, and the contents are heated on a hot plate set at a temperature that will allow a gentle reflux to occur. Heating is continued, with the addition of 2 mL increments of nitric acid as needed, until digestion of the sample is complete. Then the watch glass is removed and heating is continued until dense white fumes of SO_3 are evolved. The contents of the beaker are cooled, treated with 1 mL of (1+1) nitric acid, and diluted to 100 mL with high purity water.

Prior to aspiration into a rich nitrous oxide–acetylene flame, 2 mL of potassium chloride solution (95 g KCl/L) is added to each sample and standard. Atomic absorbances are measured at 365.3 nm. Quantification of titanium is by direct comparison.

A single laboratory using mixed industrial/domestic effluent spiked at titanium concentrations of 2, 10, and 50 mg/L obtained titanium recoveries of 97, 91, and 88%, respectively. The mean results showed coefficients of variation of 4, 1, and 0.8%, respectively.

6.37.2 Determination of Titanium; US EPA Method 283.2

US EPA Method 283.2 is applicable to the determination of titanium in wastewater samples at concentrations ranging from 50 to 500 μg/L. Samples are prepared by the procedure described in Section 6.37.1. Before titanium can be quantified by direct comparison, it is necessary to demonstrate for every sample matrix analyzed, that the method of standard additions is not needed. The absorbances of 20 μL injections of samples and samples spiked with titanium standards are measured in replicate at 365.3 nm using the following time–temperature program: dry, 30 seconds at 125°C; ash, 30 seconds at 1400°C; atomize, 15 seconds at 2800°C. Background corrections, continuous argon purge flow, and pyrolytic graphite tubes are recommended.

6.37.3 Other Procedures for Titanium

SM 3111D may be used in addition to US EPA Methods 283.1 and 283.2 for the determination of titanium in water and wastewater.

6.38 PROCEDURES FOR VANADIUM

Vanadium is determined in a rich nitrous oxide–acetylene flame using the triplet at 318.3-318.4-318.5 nm. The sensitivity is poorer than that for many other metals. Aluminum enhances the sensitivity, and corrections for this effect are made by adding aluminum to both samples and standards. Some alternate wavelengths (306.6, 306.0, 305.6, 320.2, 390.2 nm) are likely to show spectral interferences with titanium lines. Even though the thermal stability of vanadium compounds allows ashing temperatures as high as 1600°C, vanadium must be quantified by standard addition because of interferences.

6.38.1 Determination of Vanadium; US EPA Method 286.1

US EPA Method 286.1 describes the determination of vanadium in wastewater by flame atomization–atomic absorption spectrometry at concentrations ranging from 2 to 100 mg/L. The samples are prepared by the procedures described under "Water and Wastewater Samples" in Section 3.1.2.2. Prepared samples and standards are treated with aluminum nitrate solution [139 g Al $(NO_3)_3 \cdot 9H_2O$/200 mL] at a rate of 2 mL per 100 mL of sample or standard prior to aspiration into a rich nitrous oxide–acetylene flame. The atomic absorbances are measured at the 318.4 nm, and vanadium is quantified by direct comparison.

In a single laboratory using mixed industrial/domestic effluent spiked with vanadium at 2, 10, and 50 mg/L, recoveries of 100, 95, and 97%, respectively, were obtained. The coefficients of variation for the corresponding mean results were 5, 1, and 0.4%.

6.38.2 Determination of Vanadium; US EPA Method 286.2

U.S. EPA Method 286.2 describes the determination of vanadium in wastewater by electrothermal atomization–atomic absorption spectrometry. The method is applicable to the determination of vanadium in the range of 10–200 μg/L. The samples are prepared by the procedures described in Section 3.1.2.2. The vanadium concentrations of the samples are determined by the method of standard additions unless it can be demonstrated that there are no interferences, in which case direct comparison may be used. The absorbances of replicate 20 μL injections of sample solutions and of samples spiked with known amounts of vanadium are measured at 318.4 nm using the following time–temperature program: dry, 30 seconds at 125°C, ash, 30 seconds at 1400°C; atomize, 15 seconds at 2800°C. Background corrections, continuous flow argon purge, and pyrolytic graphite tubes are recommended.

6.38.3 Other Procedures for Vanadium

SM 3111D and SM 3120B may be used for the determination of vanadium in water and wastewater in addition to US EPA Methods 286.1, 286.2, and 200.7.

6.39 PROCEDURES FOR ZINC

Zinc is conveniently determined in the air–acetylene flame at a wavelength of 213.9 nm. Use of the nitrous oxide–acetylene flame results in a threefold loss in sensitivity. The volatility of some zinc compounds limits the temperature of the ashing step to 400°C. Hence, matrix interferences are frequently encountered with electrothermal atomization, and the method of standard additions is frequently used to minimize their effects.

6.39.1 Determination of Zinc; US EPA Method 289.1

US EPA Method 289.1 is applicable to the determination of zinc in wastewater samples at concentrations ranging from 0.05 to 1 mg/L. Samples are prepared by the procedures described under "Water and Wastewater Samples" in Section 3.2.2.2. For samples containing zinc below 0.05 mg/L, the chelation–extraction procedure from Section 3.2.3 may be used for preconcentration. When chelation–extraction is employed, both the samples and the standards must be carried through the procedure.

Samples and standards are aspirated into the air–acetylene flame, and the atomic absorbances at 213.9 nm are measured. The zinc concentrations of the samples are determined by direct comparison using the absorbances of the standards for calibration.

Some seven dozen laboratories analyzed natural water spiked with six synthetic concentrates containing varying amounts of aluminum, calcium, chromium, copper, iron, manganese, lead, and zinc. For samples spiked with zinc at 300 μg/L, the mean results showed a 33% coefficient of variation and a bias of 1%. The mean results for samples spiked with 60 μg Zn/L were within 10% of the theoretical value and showed a coefficient of variation of 44%.

6.39.2 Determination of Zinc; US EPA Method 289.2

The determination of zinc in wastewater at concentrations between 0.2 and 4 μg/L by electrothermal atomization–atomic absorption spectrometry is described in US EPA Method 289.2. Samples are prepared in accord with the procedures described in section 3.1.2.2, and the absorbances from replicate 20 μL injections of prepared samples and samples spiked with zinc standards are measured at 213.9 nm using the following time–temperature program: dry, 30 seconds at 125°C; ash, 30 seconds at 400°C; atomize, 10 seconds at 2500°C. If it can be shown that the method of standard addition is not needed, the concentrations of zinc in the samples may be determined by direct comparison. Background corrections, continuous argon purge, and nonpyrolytic graphite tubes are recommended.

6.39.3 Other Procedures for Zinc

Recent amendments to the Clean Water Act allow the determination of arsenic by SM 3113B, SM 3120B, and ASTM D1691-90 in addition to US EPA Methods 289.1, 289.2, 200.7, and 200.8.

7

METHODS FOR COMPLIANCE MONITORING OF LIQUID WASTES, SOLID WASTES, SLUDGES, SEDIMENTS, AND SOILS

The quantification of toxic metals in solid and liquid wastes is required for the assessment of potential hazards to public health and environmental quality. Sludges, sediments, and soils are subjected to similar scrutiny, and such measurements are required, as well, to determine whether remediation is required or completed. While many of the methods for solid waste compliance monitoring are found in "SW-846,"[40] other sources[168] contain valuable information on the applications of atomic spectrometry to regulatory compliance monitoring.

7.1 INDUCTIVELY COUPLED PLASMA–ATOMIC EMISSION SPECTROMETRY, US EPA METHOD 6010

US EPA Method 6010 is applicable to the determination of aluminum, antimony, arsenic, barium, beryllium, boron, cadmium, calcium, chromium, cobalt, copper, iron, lead, magnesium, manganese, molybdenum, nickel, potassium, selenium, silicon, silver, sodium, thallium, vanadium, and zinc in solid and liquid wastes. Samples are prepared in accordance with one of the following methods.

1. US EPA Method 3005 for total recoverable metals in surface and groundwater samples is essentially the aqua regia digestion described under "Water and Wastewater Samples" in Section 3.1.2.2 (US EPA Method 200.2): a 100 mL aliquot is treated with 2 mL of nitric acid and 5 mL of hydrochloric acid, heated on a 90–95°C steam bath or hot plate until the volume is reduced to 15–20 mL, cooled, transferred (with filtration if necessary) to a 100 mL volumetric flask, and brought to volume with ASTM Type II water.
2. US EPA Method 3010 is a nitric acid digestion for preparing aqueous sam-

ples and extracts prior to the determination of total metal by flame atomization–atomic absorption spectrometry or by inductively coupled plasma–atomic emission spectrometry: a 100 mL aliquot is treated with 3 mL of nitric acid, heated without boiling on a hot plate until the volume is reduced to 5 mL, cooled, treated with a second 3 mL of nitric acid, reheated with gentle reflux action until the volume is reduced to 5 mL, cooled, treated with 10 mL of (1+1) hydrochloric acid, warmed for 15 minutes, cooled, transferred (with filtration if necessary) to a 100 mL volumetric flask, and brought to volume with ASTM Type II water.

3. US EPA Method 3015 is a microwave-assisted nitric acid digestion of aqueous samples and extracts for the determination of metals and metalloids by four spectrometric techniques: flame atomization–atomic absorption, electrothermal atomization–atomic absorption, inductively coupled plasma–atomic emission, and inductively coupled plasma–mass spectrometry. A 45 mL aliquot is digested with 5 mL of nitric acid in a PTFE vessel for 20 minutes in a calibrated microwave unit as described in Section 3.1.2.2.
4. US EPA Method 3020 is applicable to the preparation of aqueous samples and extracts prior to the determination of total metals by electrothermal atomization–atomic absorption spectrometry. It is similar to US EPA Method 3010 but avoids the use of hydrochloric acid, which may interfere with the electrothermal atomization process. A 100 mL aliquot is treated with 3 mL of nitric acid, heated without boiling on a hot plate until the volume is reduced to 5 mL, cooled, treated with a second 3 mL portion of nitric acid, reheated with gentle reflux action until the volume is reduced to 3 mL, cooled, treated with 10 mL of ASTM Type II water, warmed for 10–15 minutes, transferred (with filtration or centrifugation if needed) to a 100 mL volumetric flask, and brought to volume with Type II water.
5. US EPA Method 3040 describes a dissolution procedure for samples containing oils, greases, and waxes prior to the quantification of metals by atomic absorption spectrometry or inductively coupled plasma–atomic emission spectrometry. US EPA Method 3040 is an alternative to the ASTM designation D 5198 mentioned under "Oily Wastes" in Section 3.1.2.2. By US EPA Method 3040, samples are diluted with xylene or sometimes 4-methyl, 2-pentanone (MIBK) to 10% (m/v). The method of standard additions with organometallic standards and background is required for the spectrometry.
6. US EPA Method 3050 is applicable to the preparation of sludge, sediment, and soil samples for atomic absorption spectrometry and for inductively coupled plasma–atomic emission spectrometry. Details on this nitric acid–hydrogen peroxide digestion are given under "Solid Waste, Sludge, Sediment, and Soil Samples" in Section 3.1.2.2.
7. US EPA Method 3051 describes a microwave-assisted, nitric acid digestion applicable to preparing sludge, sediment, and soil samples for determination of the metals contents by flame atomization–atomic absorption spectrometry, electrothermal atomization–atomic absorption spectrometry, inductively cou-

pled plasma–atomic emission spectrometry, and inductively coupled plasma–mass spectrometry. The details of this method are presented in the appropriate subsection of Section 3.1.2.2.

Method 6010 describes the quantification by inductively coupled plasma–atomic emission spectrometry of the elements listed at the beginning of this section. The measurements may be made either simultaneously or sequentially. Background corrections are necessary. Split and spike techniques (Section 1.1.4.4.) as well as the method of standard additions (Section 1.1.5.2) should be employed to confirm the absence of matrix interferences. Recommended wavelengths and estimated detection limits are listed in Table 7.1, and analyte concentration equivalent interferences are listed in Table 7.2.

The mean results obtained by Method 6010 in a seven-laboratory intercomparison for determining 14 elements at concentrations ranging from 50 to 500 μg/L were within ±10% of the theoretical values for aluminum, arsenic, beryllium, cadmium, chromium, cobalt, copper, iron, lead, manganese, nickel, vanadium, and zinc. The mean results for selenium showed a negative bias of 20%.

7.2 DETERMINATION OF TRACE ELEMENTS IN WATERS AND WASTES BY INDUCTIVELY COUPLED PLASMA–MASS SPECTROMETRY; US EPA METHOD 200.8

US EPA Method 200.8 can be used to determine aluminum, antimony, arsenic, barium, beryllium, cadmium, chromium, cobalt, copper, lead, manganese, mercury,

TABLE 7.1 Recommended Wavelengths and Estimated Detection Limits for US EPA Method 6010

Element	Wavelength (nm)	EDL (μg/L)	Element	Wavelength (nm)	EDL (μg/L)
Aluminum	308.125	45	Lead	220.353	42
Antimony	206.833	32	Magnesium	279.079	30
Arsenic	193.696	53	Manganese	257.610	2
Barium	455.404	2	Molybdenum	202.030	8
Beryllium	313.042	0.3	Nickel	231.604	15
Boron	249.773	5	Potassium	766.491	Variable
Cadmium	226.502	4	Selenium	196.026	75
Calcium	317.933	10	Silicon	288.158	5.8
Chromium	267.716	7	Silver	328.068	7
Cobalt	228.616	7	Sodium	588.995	29
Copper	324.754	6	Thallium	190.864	40
Iron	259.940	7	Vanadium	292.402	8
			Zinc	213.856	2

TABLE 7.2 Analyte Concentration Equivalents[a] Arising from Interferences at the 100 mg/L Level

	Interferants[b]									
Analyte	Al	Ca	Cr	Cu	Fe	Mg	Mn	Ni	Tl	V
Aluminum	*	*	*	*	*	*	0.21	*	*	1.4
Antimony	0.47	*	2.9	*	0.08	*	*	*	0.25	0.45
Arsenic	1.3	*	0.44	*	*	*	*	*	*	1.1
Barium	*	*	*	*	*	*	*	*	*	*
Beryllium	*	*	*	*	*	*	*	*	0.04	0.05
Boron	0.04	*	*	*	0.32	*	*	*	*	*
Cadmium	*	*	*	*	0.03	*	*	0.02	*	*
Calcium	*	*	0.08	*	0.01	0.01	0.04	*	0.03	0.03
Chromium	*	*	*	0.003	*	0.04	*	*	0.04	
Cobalt	*	*	0.03	*	0.005	*	*	0.03	0.15	*
Copper	*	*	*	*	0.003	*	*	*	0.05	0.02
Iron	*	*	*	*	*	*	0.12	*	*	*
Lead	0.17	*	*	*	*	*	*	*	*	*
Magnesium	*	0.02	0.11	*	0.13	*	025	*	0.07	0.12
Manganese	0.005	*	0.01	*	0.002	0.002	*	*	*	*
Molybdenum	0.05	*	*	*	0.03	*	*	*	*	*
Nickel	*	*	*	*	*	*	*	*	*	*
Selenium	0.23	*	*	*	0.09	*	*	*	*	*
Silicon	*	*	0.07	*	*	*	*	*	*	0.01
Sodium	*	*	*	*	*	*	*	*	0.08	*
Thallium	0.30	*	*	*	*	*	*	*	*	*
Vanadium	*	*	0.05	*	0.005*	*	*	*	0.02	*
Zinc	*	*	*	0.14	*	*	*	0.29	*	*

[a]Apparent concentration of analyte due to interferant (mg/L).
[b]Asterisk (*) indicates no interference was observed even when interferants were introduced at the following milligram levels: Al = 1000, Ca = 1000, Cr = 200, Cu = 200, Fe = 1000, Mg = 1000, Mn = 200, Tl = 200, V = 200.

molybdenum, nickel, selenium, silver, thallium, thorium, uranium, vanadium, and zinc in groundwaters, surface waters, and drinking waters, as well as in wastewaters, sludges, and solid wastes. Details on Method 200.8 are presented in Section 6.2.

7.3 DETERMINATION OF TRACE ELEMENTS BY STABILIZED TEMPERATURE GRAPHITE FURNACE ATOMIC ABSORPTION; US EPA METHOD 200.9

In addition to its applicability to the determination of trace elements in drinking water, groundwater, surface water, and wastewater, US EPA Method 200.9 can be used to determine aluminum, antimony, arsenic, beryllium, cadmium, chromium, cobalt, copper, iron, lead, manganese, nickel, selenium, silver, thallium, tin, and zinc in solid wastes, sludges, and sediments. Details on Method 200.9 are presented in Section 6.3.

7.4 ATOMIC ABSORPTION METHODS; US EPA METHOD 7000

US EPA Method 7000 presents an overview of the theoretical and practical considerations for determining metals in water, wastewater, liquid wastes, solid wastes, sludges, sediments, and soils by atomic absorption spectrometry. It is a valuable introduction to the "7000 series" methods described in Sections 7.5–7.30. Much of the information found in US EPA Method 7000 is contained in Section 1.1. Detailed descriptions of procedures for the preparation of samples of these substances prior to the quantification of metals by atomic absorption spectrometry can be found in the appropriate subsections of Section 3.1.2.2. Section 7.1 presents similar information in condensed form.

7.5 PROCEDURES FOR ALUMINUM

Section 6.4 contains additional information for the determination of aluminum.

7.5.1. Determination of Aluminum; US EPA Method 7020

US EPA Method 7020 describes the determination of aluminum by atomic absorption spectrometry in the rich nitrous oxide–acetylene flame at 309.3 nm. Background correction is not necessary; the use of a 100 ppm potassium (from KCl) ionization suppressor is recommended.

7.6 PROCEDURES FOR ANTIMONY

Section 6.5 contains additional information for the determination of antimony.

7.6.1 Determination of Antimony; US EPA Method 7040

US EPA Method 7040 is used to determine antimony concentrations by atomic absorption spectrometry with background correction in a lean air–acetylene flame at 217.6 nm. Sample preparation by US EPA Method 3005 (Section 7.1) is preferred. Antimony recoveries from samples prepared by US EPA Method 3005 were better than those reported from samples prepared by other methods, although recoveries of 96 and 97% were reported from wastewater samples prepared by US EPA Method 3010.

Spectral interferences by high concentrations of lead may necessitate making the antimony measurements at an alternate wavelength of 231.1 nm. Matrix interferences from differences in acid composition or concentration or from high concentrations of copper and nickel can be eliminated in the nitrous oxide–acetylene flame or compensated for by matching sample and standard matrices.

7.6.2 Determination of Antimony; US EPA Method 7041

US EPA Method 7041 describes the determination of antimony by electrothermal atomization–atomic absorption spectrometry. The measurements are made at 217.6 nm using a dry–ash–atomize sequence of 30 seconds at 125°C, 30 seconds at 800°C, and 10 seconds at 2700°C. Nonpyrolytic graphite tubes and uninterrupted purge gas (argon or nitrogen) flow are recommended. Background correction is required for these measurements. As indicated in Section 7.6.1, high concentrations of lead may necessitate the use of an alternate wavelength of 231.1 nm for making antimony measurements. US EPA Method 3005 (Section 7.1) is recommended for preparation of samples prior to the determination of antimony by electrothermal atomization–atomic absorption spectrometry.

7.7 PROCEDURES FOR ARSENIC

Section 6.6 contains additional information for the determination of arsenic.

7.7.1 Determination of Arsenic; US EPA Method 7060

After appropriate sample preparation, US EPA Method 7060 is applicable to the determination of arsenic in mobility (TCLP) extract, groundwater, waste and soil. This electrothermal atomization–atomic absorption spectrometry method requires background correction and the use of a nickel nitrate matrix modifier. A dry–ash–atomize sequence of 30 seconds at 125°C, 30 seconds at 1100°C, and 10 seconds at 2700°C is recommended for measurements at 193.7 nm.

Water samples are prepared by US EPA Method 3005 (aqua regia digestion), and sludge samples are prepared by US EPA Method 3050 (nitric acid–hydrogen peroxide digestion). A 5 mL aliquot of the digestate is transferred to a 10 mL volumetric flask, treated with 1 mL of nickel nitrate matrix modifier [5 g $Ni(NO_3)_2 \cdot 6H_2O$/100 mL], and brought to volume with ASTM Type II water. This modified solution is transferred into the electrothermal atomizer in 20 μL injections, and the atomic absorptions are measured by means of the dry–ash–atomize sequence just recommended. Also recommended are the use of a electrodeless discharge lamp, Zeeman background correction, nonpyrolytic graphite tubes, and argon purge gas.

US EPA Method 7060 has been applied to the determination of arsenic in samples of emission control dust, contaminated soil, oily soil, and estuarian sediment (NIST SRM 1646) prepared by US EPA Method 3050. Arsenic recovery from the estuarian sediment was 70–75%.

7.7.2 Determination of Arsenic: US EPA Method 7061

US EPA Method 7061 describes the quantification of arsenic by atomic absorption spectrometry of the gaseous hydride. Waste samples prepared by the methods described in Section 7.1 are subjected to further digestion with nitric and sulfuric acids prior to making the atomic absorption measurements.

A 50 mL aliquot of prepared sample is transferred to a 100 mL beaker, treated with 10 mL of nitric acid and 12 mL of sulfuric acid, and heated on a hot plate until the volume is reduced to 20 mL and dense white fumes of SO_3 are evolved. During this treatment, additional 3 mL increments of nitric acid may be added to maintain oxidizing conditions (i.e., to prevent charring or darkening of the sample). When a colorless or pale yellow solution is obtained, heating is stopped, and the contents of the beaker are allowed to cool. To complete the treatment, the contents of the beaker are treated with 25 mL of ASTM Type II water and again heated until dense white fumes of SO_3 are evolved. Then the contents of the beaker are cooled, transferred to a 100 mL volumetric flask, treated with 40 mL of hydrochloric acid, and brought to volume with Type II water.

A 25 mL aliquot of the treated sample is transferred to the reaction vessel of the hydride generator, connected to an argon purge supply, and treated with 1 mL of potassium iodide solution (20 g in 100 mL water) and 0.5 mL of tin(II) chloride solution (100 g in 100 mL of hydrochloric acid). The reaction vessel is allowed to stand for 10 minutes before being injected with 1.5 mL of zinc slurry. Injection of the zinc slurry immediately produces arsine, which is swept into the hydrogen–argon flame for atomization and measurement of atomic absorption at 193.7 nm with background correction.

Compensation for potential interferences from high concentration of chromium, cobalt, copper, mercury, molybdenum, nickel, and silver and trace residues of nitric acid can be made with the method of standard addition. The method of standard addition is required when TCLP extracts are examined.

7.7.3 Determination of Arsenic in Wastes from Nonferrous Smelters; US EPA Method 108

Wastes from primary copper refineries and other nonferrous smelters may contain significant quantities of arsenic. US EPA Methods 108A and 108B are applicable to the determination of arsenic in samples of process ores and reverberatory furnace mattes.[169]

Samples are collected and homogenized using the procedures described in Section 2.2.5. Representative specimens are decomposed with acids in PTFE vessels and analyzed for arsenic by atomic absorption spectrometry (flame, gaseous hydride, or electrothermal atomization).

For Method 108A, representative specimens of ore or matte in the mass range of 50–500 mg are weighed into the PTFE cups of steel-jacketed digestion vessels[170] and treated with 2 mL of hydrofluoric acid and 2 mL of nitric acid. The vessels are immediately sealed to prevent losses by volatilization and placed in a 105°C oven for 2 hours. After this time, the digestion vessels are removed from the oven, allowed to cool to room temperature, and carefully opened. The contents of the PTFE cups are filtered through PTFE filters into 50 mL polypropylene volumetric flasks. The cups are rinsed three times with small amounts of 0.5 N nitric acid, and the washings are added through the filters to the corresponding flasks. The content of the flasks are treated with 5 mL of solution containing 10% (m/v) potassium chloride in 3% (v/v) nitric acid and brought to volume with 0.5 N nitric acid.

For Method 108B, representative specimens of ore or matte in the mass range of 100–1000 mg are weighed into the PTFE beakers and treated with 15 mL of nitric acid and 10 mL each of hydrochloric, hydrofluoric, and perchloric acids, in the order listed. After being allowed to stand for 10 minutes in an approved fume hood, the contents of the beakers are heated on a hot plate until the volumes are reduced to 2–3 mL. Following cooling to room temperature, the contents of the beakers are diluted with 20 mL of water and 10 mL of hydrochloric acid and warmed to completely dissolve the digested material. The contents of the beakers are then cooled to room temperature and transferred to 100 mL volumetric flasks and brought to volume with water.

Flame atomization–atomic absorption spectrometry may be used for quantification when the concentration is greater than 10 μg/mL. For arsenic concentrations below 10 μg/mL, gaseous hydride or electrothermal atomization–atomic absorption spectrometry should be employed.

For the quantification of arsenic by atomic absorption in the air–acetylene or argon–hydrogen flame, measurements of samples and standards are made at 193 nm using an electrodeless discharge lamp as the source of resonance radiation.

Additional sample preparation is required when quantification of arsenic is by gaseous hydride atomization–atomic absorption spectrometry. An aliquot of solution prepared by either Method 108A or by Method 108B and containing about 1–5 μg of arsenic is transferred to the hydride generator reaction vessel, where it is diluted with 15 mL of water. After additions of 15 mL of concentrated hydrochloric acid followed by 15 mL of 30% (m/v) potassium iodide solution have been made to the reaction vessel, this receptacle is heated in a 50°C water bath for 5 minutes. The reaction vessel is cooled and connected to the hydride generator. When the spectrometer, which has been fitted with an electrodeless discharge lamp for arsenic and a heated quartz cell for atomizing arsine, has stabilized at baseline, 5 mL of 5% sodium borohydride in 0.1 N sodium hydroxide solution is injected into the hydride generator, and the atomic absorbance of arsenic at 193 nm is recorded for 30 seconds. Quantification is achieved by direct comparison of absorbances from the samples to those from arsenic standards.

Quantification of arsenic by electrothermal atomization–atomic absorption spectrometry requires the presence of nickel nitrate and hydrogen peroxide modifiers. To a 5 mL aliquot of solution prepared by Method 108A or 108B and contained in a 10 mL volumetric flask are added 1 mL of 1% (m/v) nickel nitrate in 50% (v/v) nitric acid and 1 mL of 3% hydrogen peroxide. The contents of the flask are brought to volume with water. Arsenic is quantified by comparison of absorbances at 193 nm from standards and samples injected into the electrothermal atomizer.

7.8 PROCEDURES FOR BARIUM

Section 6.7 contains additional information for the determination of barium.

7.8.1 Determination of Barium; US EPA Method 7080

US EPA Method 7080 describes the determination of barium by atomic absorption spectrometry in the rich nitrous oxide–acetylene flame at 553.6 nm. Background correction is not required. Potassium ionization suppressor (95 g KCl/L) must be added at a rate of 2 mL per 100 mL sample or standard. Quantification is by direct comparison.

Single-laboratory precision for the determination of barium in wastewater samples prepared by US EPA Method 3010 was ±10%, and spike recoveries ranged from 94 to 113% for samples containing between a few tenths and a few ppm of barium.

7.9 PROCEDURES FOR BERYLLIUM

Additional information for the determination of beryllium is contained in Section 6.8.

7.9.1 Determination of Beryllium; US EPA Method 7090

In US EPA Method 7090, the concentration of beryllium in prepared waste samples is determined by atomic absorption spectrometry in a rich nitrous oxide–acetylene flame at 234.9 nm. Corrections for background are required. When interferences by high concentrations of aluminum are encountered, they may be eliminated by the addition of 1% fluoride. The addition should also be made to the standards when quantification is by direct comparison. Standard addition may be employed to compensate for interferences from magnesium and silicon at high concentrations.

Single-laboratory precision for the determination of beryllium in wastewater samples prepared by US EPA Method 3010 was better than ±10%, and spike recoveries ranged from 97 to 100% for samples containing amounts of beryllium ranging from a few hundredths to a few tenths ppm.

7.9.2 Determination of Beryllium; US EPA Method 7091

US EPA Method 7091 describes the determination of beryllium by electrothermal atomization–atomic absorption spectrometry. Using a dry–ash–atomize sequence of 30 seconds at 125°C, 30 seconds at 100°C, and 10 seconds at 2800°C, the measurements are made in nonpyrolytic graphite tubes at 234.9 nm with uninterrupted argon purge gas flow. Background correction is required for these measurements.

7.10 PROCEDURES FOR CADMIUM

Additional information for the determination of cadmium is contained in Section 6.10.

7.10.1 Determination of Cadmium; US EPA Method 7130

US EPA Method 7130 describes the determination of cadmium concentrations in prepared waste samples by atomic absorption spectrometry in a lean air–acetylene flame at 228.8 nm. Background corrections are required. Cadmium is quantified by direct comparison. US EPA Method 7130 has been applied to the determination of cadmium in samples of emission control dust and wastewater treatment sludge prepared by US EPA Method 3050.

7.10.2 Determination of Cadmium; US EPA Method 7131

US EPA Method 7131 describes the determination of cadmium by electrothermal atomization–atomic absorption spectrometry. The atomic absorption measurements are made on 20 μL injections in nonpyrolytic graphite tubes at 228.8 nm with uninterrupted argon purge gas flow using a dry–ash–atomize sequence of 30 seconds at 125°C, 30 seconds at 500°C, and 10 seconds at 1900°C. Background corrections are required for these measurements.

US EPA Method 7131 has been applied to the determination of cadmium in samples of lagoon soil, oily waste, and estuarian sediment (NIST SRM 1646) prepared by US EPA Method 3050. Cadmium recovery from the estuarian sediment was exceeded 95%.

7.11 PROCEDURES FOR CALCIUM

Section 6.11 contains additional information for the determination of calcium.

7.11.1 Determination of Calcium; US EPA Method 7140

US EPA Method 7140 describes the determination of calcium by atomic absorption spectrometry in a stoichiometric nitrous oxide–acetylene flame at 422.7 nm. Background correction is not necessary. A lanthanum chloride releaser must be added to overcome chemical interferences from phosphate, borate, silicate, chromate, sulfate, vanadate, titanate, and aluminate. The lanthanum chloride releaser is prepared by dissolving 29 g of La_2O_3 in 250 mL of hydrochloric acid and diluting the resulting solution with ASTM Type II water to a final volume of 500 mL. The releaser is used at a rate of 2 mL per 100 mL sample or standard, and the calcium is quantified by direct comparison.

7.12 PROCEDURES FOR CHROMIUM

Additional information for the determination of chromium is contained in Section 6.12.

7.12.1 Determination of Chromium; US EPA Method 7190

The determination chromium concentrations in prepared waste samples by atomic absorption spectrometry is described in US EPA Method 7190. Measurements are made in a rich nitrous oxide–acetylene flame at 357.9 nm. Background corrections are not required. The concentration of chromium in the samples is quantified by direct comparison of the absorbances to those of chromium standards. US EPA Method 7190 has been applied to the determination of chromium in samples of emission control dust and wastewater treatment sludge prepared by US EPA Method 3050.

7.12.2 Determination of Chromium; US EPA Method 7191

US EPA Method 7191 describes the determination of chromium by electrothermal atomization–atomic absorption spectrometry. Atomic absorption measurements are made in nonpyrolytic graphite tubes at 357.9 nm with uninterrupted argon purge gas flow using a dry–ash–atomize sequence of 30 seconds at 125°C, 30 seconds at 1000°C, and 10 seconds at 2700°C. Background corrections are not required for these measurements.

US EPA Method 7191 has been applied to the determination of chromium in samples, prepared by US EPA Method 3050, of used lubricating oil (NIST SRM 1085), oily lagoon soil, contaminated soil, EPA quality control sludge, estuarian sediment (NIST SRM 1646), and paint primer. Chromium recovery from the estuarian sediment was only slightly above 50%. Chromium recovery from the EPA QC sludge was slightly over 75%. Others[109] have observed similar recoveries of chromium from this material. Chromium recoveries from SRM 1085 were from 104 to 119% of the certified mean.

7.12.3 Determination of Chromium; US EPA Method 7195

US EPA Method 7195 makes use of the lead sulfate coprecipitation procedure mentioned in Section 3.2.2 to separate hexavalent chromium from trivalent chromium prior to quantification of the former by either flame (see Section 7.12.1) or electrothermal (see Section 7.12.2) atomization atomic absorption spectrometry.

A portion of the TCLP extract is adjusted to pH 3.5 ± 0.3 by dropwise addition of (1+9) acetic acid or (1+9) ammonia solution. After pH adjustment, a 10 mL aliquot is transferred to a centrifuge tube, treated with 100 μL of lead nitrate solution [33.1 g of Pb $(NO_3)_2$ dissolved in and diluted to 100 mL with ASTM Type II water], and allowed to stand for 3 minutes. The contents of the centrifuge tube are then treated with 0.5 mL of glacial acetic acid and 100 μL of ammonium sulfate solution [2.7 g of $(NH_4)_2SO_4$ dissolved in and diluted to 100 mL with ASTM Type II water]. The contents of the centrifuge tube are stopped, mixed, and centrifuged at speeds slowly increased over 5 minutes to 2000 rpm. The supernate is removed by aspiration, and the precipitate is washed twice with 5 mL of Type II water. The washed precipitate is dissolved in 0.5 mL of nitric acid. Finally, 100 μL of 30% hydrogen peroxide and 100 μL of calcium nitrate solution [11.8 g of $Ca(NO_3)_2 \cdot 4H_2O$ dissolved in and diluted to 100 mL with Type II water] are added, and the contents of the centrifuge tube are diluted to 10 mL with Type II water. The concentration of chromium in the resulting solution is determined by means of standard additions carried through the procedure using either flame or electrothermal atomic–absorption spectrometry as described earlier.

7.12.4 Determination of Chromium; US EPA Method 7197

US EPA Method 7197 makes use of the APDC-MIBK chelation–extraction procedure mentioned in Section 3.2.3 to separate hexavalent chromium from trivalent chromium prior to quantification of the former in the organic phase by flame atomization–atomic absorption spectrometry at 357.9 nm.

An aliquot of aqueous waste or extract containing no more than 2.5 μg of chromium is diluted to 100 mL, adjusted to pH 2.4 with 0.12 M sulfuric acid solution and/or 1 M sodium hydroxide solution, and transferred to a 200 mL volumetric flask. The contents of the flask are treated with 5 mL of 1% (m/v) aqueous APDC solution and mixed. After addition of 10 mL of MIBK, the contents of the flask are shaken vigorously for 3 minutes. The phases are allowed to separate, and then water is added to float the organic phase high into the neck of the flask. The reagent blank and the hexavalent chromium standards are prepared similarly. The organic phases are aspirated into a rich air–acetylene flame, and the hexavalent chromium content of the sample is quantified by direct comparison to the standards. For TCLP extracts, quantification must be by standard additions.

7.13 PROCEDURES FOR COBALT

Additional information for the determination of cobalt is contained in Section 6.13.

7.13.1 Determination of Cobalt; US EPA Method 7200

US EPA Method 7200 describes the determination cobalt concentrations in prepared waste samples by atomic absorption spectrometry in a lean air–acetylene flame at 240.7 nm. Background corrections are required. Cobalt is quantified by direct comparison. US EPA Method 7200 has been applied to the determination of cobalt in samples of wastewater. At concentrations of 1 and 5 ppm, precision was ±10% and cobalt recovery was 97%.

7.13.2 Determination of Cobalt; US EPA Method 7201

US EPA Method 7201 describes the determination of cobalt by electrothermal atomization–atomic absorption spectrometry. The atomic absorption measurements are made on 20 μL injections in nonpyrolytic graphite tubes at 240.7 nm with uninterrupted argon purge gas flow using a dry–ash–atomize sequence of 30 seconds at 125°C, 30 seconds at 900°C, and 10 seconds at 2700°C. Background corrections are required for these measurements.

7.14 PROCEDURES FOR COPPER

Section 6.14 contains additional information for the determination of copper.

7.14.1 Determination of Copper; US EPA Method 7210

Method 7210 describes the determination of copper by atomic absorption spectrometry in a lean air–acetylene flame at 324.7 nm. Background correction is recommended.

7.15 PROCEDURES FOR IRON

Section 6.17 contains additional information for the determination of iron.

7.15.1 Determination of Iron; US EPA Method 7380

US EPA Method 7380 describes the determination of iron by atomic absorption spectrometry in the lean air–acetylene flame at 248.3 nm. Background correction is required, and matrix matching is recommended.

7.16 PROCEDURES FOR LEAD

Additional information on the determination of lead is contained in Section 6.18.

7.16.1 Determination of Lead: US EPA Method 7420

This method describes the determination of lead by flame atomization atomic absorption spectrometry. Samples and standards are aspirated into a lean, blue air-acetylene flame, and absorbances are measured at 283.3 nm. An alternate wavelength for measuring the atomic absorbances is 217.0 nm. Background correction is required. The working range of the method is from 1 to 20 mg/L. For the determination of lead at concentrations below 0.2 mg/L, the furnace technique, US EPA Method 7421, is required.

7.16.2 Determination of Lead: US EPA Method 7421

Method 7421 is applicable to the determination of lead by electrothermal atomization atomic absorption spectrometry in the concentration range from 5 to 100 μg/L. Measurements are made at 283.3 nm using argon purge gas and a dry-ash-atomize cycle of 30 seconds at 125°C, 30 seconds at 500°C, and 10 seconds at 2700°C. Background correction is required.

7.17 PROCEDURES FOR MAGNESIUM

Section 6.19 contains additional information for the determination of magnesium.

7.17.1 Determination of Magnesium; US EPA Method 7450

US EPA Method 7450 describes the determination of magnesium by atomic absorption spectrometry in a lean air–acetylene flame at 285.2 nm. Background correction is required. Addition of a lanthanum chloride releaser is necessary to overcome chemical interferences from phosphate, borate, silicate, chromate, sulfate, vanadate, titanate, and aluminate. The lanthanum chloride releaser is prepared by dissolving 29 g of La_2O_3 in 250 mL of hydrochloric acid and diluting the resulting solution with ASTM Type II water to a final volume of 500 mL. The releaser is used at a rate of 2 mL per 100 mL of sample or standard, and the magnesium is quantified by direct comparison.

Single-laboratory precision rates for wastewaters containing 2 and 8 ppm magnesium were ±0.1 and ±0.2 ppm, respectively; the recovery of magnesium was 100% in both cases.

7.18 PROCEDURES FOR MANGANESE

Additional information for the determination of manganese is contained in Section 6.20.

7.18.1 Determination of Manganese: US EPA Method 7460

The determination of manganese is described in US EPA Method 7460. The determination is made by atomic absorption spectrometry in a slightly lean to stoichiometric air–acetylene flame at 279.5 nm. Background correction is required, and quantification is by direct comparison to matrix matched standards or by standard addition.

7.19 PROCEDURES FOR MERCURY

For additional information on the atomic absorption spectrometry of mercury, see Section 6.21 and "Cold Vapor Generators" in Section 1.1.3.4.

7.19.1 Determination of Mercury in Liquid Waste; US EPA Method 7470

US EPA Method 7470 is applicable to the quantification of mercury in samples of groundwater and wastewater as well as in aqueous extracts of solid waste and soil samples. For the quantification of mercury in sludges, sediments, and soils, it is necessary to use US EPA Method 7471, described shortly (Section 7.19.2).

An aliquot of the sample solution containing less than 1 μg of mercury is diluted to 100 mL with ASTM Type II water in the reaction vessel of a cold vapor generator. With mixing after each addition, 5 mL of sulfuric acid, 2.5 mL of nitric acid, and 15 mL of 5% (m/v) potassium permanganate solution are added to the contents of the reaction vessel and allowed to incubate for at least 15 minutes at room temperature. Additional 1 mL increments of 5% potassium permanganate solution may be necessary to maintain oxidizing conditions in the reaction vessel, as shown by the persistence of the purple color, during the 15-minute incubation. (*Note*: Equal amounts of 5% potassium permanganate solution must be added to the mercury standards and reagent blanks.)

To begin, 8 mL of 5% (m/v) potassium peroxydisulfate solution is added to the reaction vessel, and the contents are incubated in a 95°C water bath for 2 hours. The contents of the reaction vessel are the cooled to room temperature and treated with a solution of 6 mL of 12% (m/v) sodium chloride and 12% (m/v) hydroxylamine sulfate.

The contents of the reaction vessel are treated with 5 mL of tin(II) sulfate suspension (25 g of $SnSO_4$ + 250 mL of 0.25 M H_2SO_4), and the vessel is immediately connected to the aeration apparatus for purging mercury vapor from the vessel and transporting mercury vapor to the cell for measurement of atomic absorption at 253.7 nm. The atomic absorption is recorded at peak height on an appropriate electronic device.

Quantification is by comparison of peak height from the waste sample with peak heights from the mercury standards and reagent blanks. Standard addition is required for quantifying mercury in TCLP extracts.

Oxidation of chloride ion to chlorine gas, which also absorbs radiation in the 250 nm region of the spectrum, may cause interference to quantification of mercury by the cold vapor method. To eliminate this potential interference, excess (25 mL) sodium chloride–hydroxylamine sulfate solution is used, and the headspace of the reaction vessel is purged prior to the injection of the tin(II) sulfate solution. Potential interference from entrained water or from water vapor is eliminated by the inclusion of a drying tube in the transfer line between the reaction vessel and the absorption cell.

7.19.2 Determination of Mercury in Solid or Semisolid Waste; US EPA Method 7471

US EPA Method 7471 is applicable to the determination of mercury in sludges, sediments, and soils. It differs from Method 7470 in the manner by which samples are prepared for cold vapor atomic absorption spectrometry.

Triplicate 200 mg specimens of sample are transferred to the cold vapor generator reaction vessels. Five mL of ASTM Type II water and 5 mL of aqua regia are added, and the contents of the vessels are heated in a 95°C water bath for 2 minutes. Upon addition of 50 mL of Type II water and 15 mL of 5% (m/v) potassium permanganate solution, the vessels are returned to the water bath for an additional 30 minutes. The contents of the reaction vessels are cooled to room temperature and treated with a solution of 6 mL of 12% (m/v) hydroxylamine sulfate–12% (m/v) sodium chloride.

As an alternative to the aqua regia digestion, the samples (200 mg) may be autoclaved in the reaction vessels covered with aluminum foil for 15 minutes at 121°C and 15 lb with 5 mL of sulfuric acid solution, 2 mL of nitric acid, and 5 mL of saturated potassium permanganate solution. The contents of the reaction vessel are then cooled to room temperature, diluted to 100 mL with Type II water, and treated with 6 mL of the 12% (m/v) hydroxylamine sulfate–12% (m/v) sodium chloride solution.

The contents of the reaction vessels from either preparation are treated individually with 5 mL of tin(II) sulfate suspension (25 g of $SnSO_4$ + 250 mL of 0.25 M H_2SO_4), and the vessel is immediately connected to the aeration apparatus for purging mercury vapor from the vessel and transporting mercury vapor to the cell for atomic absorption measurement at 253.7 nm as described in Section 7.19.1. Quantification of mercury and elimination of chloride ion interferences are also described in Section 7.19.1.

7.19.3 Determination of Mercury in Wastewater Treatment Plant Sewage Sludges; US EPA Method 105

US EPA Method 105 is applicable to the determination of total mercury in sewage sludges prior to their incineration.[171] Sludge samples are digested with aqua regia, and their mercury contents are quantified by atomic absorption spectrometry using cold vapor atomization.

A time-based composite sample of dewatered sludge is collected from the conveyor system in 1-liter increments every 30 minutes for 8 hours prior to the drying or incineration stage. The composite is mixed for 30 minutes at 30 rpm. With intermittent mixing, six 100 mL portions of sludge are combined in a 2-liter blender and blended for 5 minutes. Four 20 mL aliquots of the blended sludge are weighed into separate 125 mL Erlenmeyer flasks, one of which will be brought to constant weight in a 105°C oven for the determination of total solids.

To each of the three remaining aliquots of blended sludge is added 25 mL of aqua regia. The contents of the flasks are heated without boiling on a hot plate for

30 minutes or until pale yellow solutions are obtained. The contents of the Erlenmeyer flasks are cooled and filtered into 100 mL volumetric flasks through S & S No. 588 paper or the equivalent. The Erlenmeyer flasks and filters are washed with 50 mL of water, and the washings are added to the contents of the corresponding volumetric flasks. The contents of the volumetric flasks are diluted to 100 mL with water.

A 5 mL aliquot from the contents of the volumetric flask or a 5 mL aliquot of mercury standard is added to the reaction vessel of the cold vapor generator, which already contains 25 mL of water and 3 drops of Antifoam, a silicon emulsion available from the J. T. Baker, Inc. (Phillipsburg, NJ). A PTFE-coated stirring bar is added to the reaction vessel, followed by 5 mL of 15% (v/v) nitric acid and 5 mL of 5% (m/v) potassium permanganate solution. The reaction vessel is connected to the cold vapor generator, the outlet to the absorption cell is closed, and the contents are stirred. The contents of the reaction vessel of the cold vapor generator are treated initially with 5 mL of 12% (m/v) sodium chloride–12% (m/v) hydroxylamine hydrochloride solution. If this volume is not adequate to decolorize the contents, 1 mL increments of this solution are added until the yellow color has disappeared. Next 5 mL of 4% (m/v) tin(II) chloride in 10% (v/v) hydrochloric acid is added to the reaction vessel. The reaction vessel is closed, and its contents stirred for 15 seconds. The outlet to the absorption cell is opened, and with continued stirring of the contents, the mercury vapor is gas-purged from the reaction vessel. The mercury concentrations in the sample solutions are determined from their atomic absorptions at 254 nm relative to those of the standards.

7.19.4 Determination of Mercury in Sediment; US EPA Method 245.5

US EPA Method 245.5 is applicable to the determination of total mercury in sludges, sediments, and soils.

Samples are collected by the procedures described in Sections 2.2.3–2.2.5.

A 200 mg specimen of the sample is placed on the bottom of the reaction vessel from the cold vapor generator. In some apparatus a 300 mL BOD bottle is used for this purpose.

The contents of the vessel are treated with 5 mL of aqua regia, and the top of the vessel is closed and covered with aluminum foil or Parafilm. The contents of the vessel are incubated briefly (2 min) in a 95°C water bath, cooled, treated with 50 mL of ASTM Type II water and 15 mL of 5% (m/v) potassium permanganate solution, closed and covered as before, and returned to the 95°C water bath incubator for 30 minutes.

Alternatively, the contents of the reaction vessel are treated with 5 mL of sulfuric acid, 2 mL of nitric acid, and 5 mL of saturated potassium permanganate solution. The top of the vessel is covered with aluminum foil, and sample digestion is completed by autoclaving for 15 minutes at 121°C and 1 atm.

Contents of the reaction vessels containing the digested samples, the positive and negative controls, and the calibration standards are diluted to 75 mL with Type II water. The contents of the reaction vessels are individually prepared for cold vapor atomic absorption spectrometry.

After 6 mL of 12% (m/v) sodium chloride–12% (m/v) hydroxylamine sulfate solution has been added, the reaction vessel is connected to the cold vapor generator and the headspace is purged to remove chlorine and other gaseous interferants. Then 5 mL of 10% tin(II) chloride or tin(II) sulfate (m/v) in 0.25 M sulfuric acid is injected into the reaction vessel, and the mercury vapor is swept into the absorption cell of the spectrometer for measurement, with electronic recording of the atomic absorption at 253.7 nm.

The mercury contents of the samples are determined from the peak height of the electronic recording relative to those of the standards and corrected to a dry mass basis.

Recoveries from sediment samples spiked with methylmercury chloride and reported to have total mercury concentrations of 0.29 ± 0.02 and 0.82 ± 0.03 μg/g were 97 and 94%, respectively.

7.20 PROCEDURES FOR MOLYBDENUM

Additional information for the determination of molybdenum is contained in Section 6.22.

7.20.1 Determination of Molybdenum; US EPA Method 7480

US EPA Method 7480 describes the determination molybdenum by atomic absorption spectrometry in a rich nitrous oxide–acetylene flame at 313.3 nm. Background corrections are required. The addition of aluminum nitrate modifier [139 g of $Al(NO_3)_3 \cdot 9H_2O$ dissolved with heat in 150 mL of ASTM Type II water and diluted to a final volume of 200 mL] at a rate of 2 mL per 100 mL sample or standard is required also. Molybdenum is quantified by direct comparison.

US EPA Method 7480 has been applied to the determination of molybdenum in samples of wastewater prepared by US EPA Method 3010. Mean concentrations and the corresponding standard deviations were 0.3 ± 0.007, 1.5 ± 0.02 and 7.5 ± 0.07 mg/L, respectively. Recovery of molybdenum was 95% or better.

7.20.2 Determination of Molybdenum; US EPA Method 7481

US EPA Method 7481 describes the determination of molybdenum by electrothermal atomization–atomic absorption spectrometry. The atomic absorption measurements are made on 20 μL injections in nonpyrolytic graphite tubes at 313.3 nm with uninterrupted argon purge gas flow using a dry–ash–atomize sequence of 30 seconds at 125°C, 30 seconds at 1400°C, and 10 seconds at 2800°C. Background corrections are required for these measurements.

7.21 PROCEDURES FOR NICKEL

Additional information for the determination of nickel is contained in Section 6.23.

7.21.1 Determination of Nickel; US EPA Method 7520

US EPA Method 7520 describes the determination of nickel concentrations in prepared waste samples by atomic absorption spectrometry. The measurements are made in a lean air–acetylene flame at 232.0 nm. Background corrections are required. Interferences from iron, cobalt, and chromium can be compensated for by matrix matching. Alternatively, the 352.4 nm resonance line can be used, but sensitivity is diminished. Nickel is quantified by direct comparison. Parasitic, nonabsorbing radiation from the nickel line at 232.2 nm causes deviations from linearity in the calibration curve.

US EPA Method 7520 has been applied to the determination of nickel in samples of wastewater treatment sludge prepared by US EPA Method 3050 and in samples of wastewater prepared by US EPA Method 3010. Mean concentrations and the corresponding standard deviations for nickel in the latter were 0.2 ± 0.011, 1.0 ± 0.02 and 5.0 ± 0.04 mg/L, respectively. Recoveries of nickel from the wastewaters were 100, 97, and 93%, respectively.

7.22 PROCEDURES FOR POTASSIUM

Section 6.28 contains additional information for the determination of potassium.

7.22.1 Determination of Potassium; US EPA Method 7610

US EPA Method 7610 describes the determination of potassium by atomic absorption spectrometry in a slightly lean air–acetylene flame at 766.5 nm. Background correction is not necessary. The addition of a 1000 ppm sodium (from NaCl) ionization suppressor at a rate of 2 mL per 100 mL of sample or standard is recommended.

In a single laboratory, potassium results for wastewater samples prepared by US EPA Method 3010 were 1.6 ± 0.2 and 6.3 ± 0.5 ppm, and the corresponding recoveries were 103 and 102%.

7.23 PROCEDURES FOR SELENIUM

Section 6.31 contains additional information for the determination of selenium.

7.23.1 Determination of Selenium; US EPA Method 7740

US EPA Method 7740 is applicable to the determination of selenium after appropriate sample preparation.

Sludge samples are prepared by US EPA Method 3050.

For aqueous samples, a 100 mL aliquot is transferred to beaker and treated with 2 mL of 30% hydrogen peroxide. Sufficient nitric acid to make the concentration 1% (v/v) is added to the contents of the beaker. The contents of the beaker are heated on a 95°C hot plate until the volume is reduced to slightly less than 50 mL. The contents of the beaker are diluted to 50 mL with ASTM Type II water. A 5 mL aliquot from the contents of the beaker is transferred to a 10 mL volumetric flask, treated with 1 mL of nickel nitrate matrix modifier [5 g of $Ni(NO_3)_2 \cdot 6H_2O$/100 mL], and brought to volume with ASTM Type II water.

The atomic absorptions from 20 μL injections of the modified solution are measured with background correction at 196.0 nm from a selenium electrodeless discharge lamp in the electrothermal atomizer. A dry–ash–atomize sequence 30 seconds at 125°C, 30 seconds at 1000°C, and 10 seconds at 2000°C is recommended. Also recommended is the use of a nonpyrolytic graphite tube and argon purge gas.

US EPA Method 7740 has been applied to the determination of selenium in samples of emission control dust prepared by US EPA Method 3050.

7.23.2 Determination of Selenium: US EPA Method 7741

US EPA Method 7741 describes the quantification of selenium by atomic absorption spectrometry of the gaseous hydride at 196.0 nm. Samples prepared by the methods described in Section 7.1 are subjected to further digestion with nitric and sulfuric acids prior to making the atomic absorption measurements.

A 50 mL aliquot of prepared sample is transferred to a 100 mL beaker, treated with 10 mL of nitric acid and 12 mL of sulfuric acid, and heated on a hot plate until the volume is reduced to 20 mL and dense white fumes of SO_3 are evolved. During this treatment, additional 3 mL increments of nitric acid may be added to maintain oxidizing conditions (i.e., to prevent charring or darkening of the sample). When a colorless or pale yellow solution is obtained, heating is stopped, and the contents of the beaker are allowed to cool. To complete the treatment of the sample, the contents of the beaker are treated with 25 mL of Type II water and again heated until dense white SO_3fumes are evolved. The contents of the beaker are cooled, transferred to a 100 mL volumetric flask, treated with 40 mL of hydrochloric acid, and brought to volume with Type II water.

A 25 mL aliquot of the treated sample is transferred to the reaction vessel of the hydride generator, connected to an argon purge supply, and treated with 0.5 mL of tin(II) chloride solution (100 g in 100 mL of hydrochloric acid). The reaction vessel is allowed to stand for 10 minutes before being injected with 1.5 mL of zinc slurry. Injection of the zinc slurry immediately produces hydrogen selenide, which is swept into the hydrogen–argon flame for atomization and measurement of atomic absorption at 196.0 nm with background correction.

Quantification is by comparison of peak height from the waste sample with peak heights from the selenium standards and reagent blanks. Standard addition is required for quantifying selenium in TCLP extracts.

7.23.3 Determination of Selenium; US EPA Method 7742

US EPA Method 7742 is applicable to the determination of selenium as the gaseous hydride; reduction of selenium to hydrogen selenide is accomplished by means of sodium borohydride.

7.24 PROCEDURES FOR SILVER

Additional information on the determination of silver is contained in Section 6.33.

7.24.1 Determination of Silver; US EPA Method 7760

US EPA Method 7760 describes the determination of silver in soils, wastes, waters, and aqueous mobility extracts after appropriate sample preparation.

A well-mixed aliquot of aqueous sample is transferred to a beaker and treated with 3 mL of nitric acid. The contents of the beaker are evaporated, to near dryness on a hot plate, without boiling. Additional 3 mL increments of nitric acid are added, and the heating continued, until a colorless or pale yellow solution is obtained; the nitric acid concentration of the final solution will be about 0.5% (v/v). The contents of the beaker are poured through a glass fiber filter into a volumetric flask of appropriate size and brought to volume with ASTM Type II water.

If silver chloride has precipitated, the contents of the beaker are made basic with ammonia solution, treated with 1 mL of cyanogen iodide solution,* and allowed to stand for 1 hour before being filtered into the volumetric flask.

Silver standards and reagent blanks must be matched to the nitric acid or to the cyanogen iodide matrix containing the sample.

Samples and standards are aspirated into a lean air–acetylene flame, and the atomic absorptions at 328.1 nm are recorded. The concentration of silver in the sample is quantified by direct comparison, or in the case of TCLP extracts, by standard addition.

US EPA Method 7760 has been applied to the determination of selenium in samples of wastewater treatment sludge and emission control dust prepared by US EPA Method 3050.

7.25 PROCEDURES FOR SODIUM

Section 6.34 contains additional information for the determination of sodium.

*Cyanogen iodide solution is prepared by dissolving 6.5 g of potassium cyanide in 50 mL of Type II water, adding 4.0 mL of ammonia solution and 5.0 mL of 1.0 N iodine solution, and diluting to a final volume of 100 mL. The cyanogen iodide solution is stable for 2 weeks.

7.25.1 Determination of Sodium; US EPA Method 7770

US EPA Method 7770 describes the determination of sodium by atomic absorption spectrometry in a lean air–acetylene flame at 589.6 mn. Background correction is not necessary. The addition of a 1000 ppm potassium (from KCl) ionization suppressor at a rate of 2 mL per 100 mL of sample or standard is recommended.

In a single laboratory, sodium results for wastewater samples prepared by US EPA Method 3010 were 8.2 ± 0.1 and 52 ± 0.8 ppm, and the corresponding recoveries were 102 and 100%.

7.26 PROCEDURES FOR THALLIUM

Additional information for the determination of thallium is contained in Section 6.35.

7.26.1 Determination of Thallium; US EPA Method 7840

US EPA Method 7840 is applicable to the determination of thallium in prepared waste samples by atomic absorption spectrometry in a lean air–acetylene flame at 276.8 nm. Corrections for background are required. Thallium is quantified by direct comparison. US EPA Method 7840 has been applied to the determination of thallium in samples of wastewater treatment prepared by US EPA Method 3010. In a single laboratory, thallium results were 0.6 ± 0.018, 3 ± 0.05, and 15 ± 0.2 ppm. Thallium recoveries were 98% or better.

7.26.2 Determination of Thallium; US EPA Method 7841

US EPA Method 7841 describes the determination of thallium by electrothermal atomization–atomic absorption spectrometry. Atomic absorption measurements are made in nonpyrolytic graphite tubes at 276.8 nm with uninterrupted argon or nitrogen purge gas flow using a dry–ash–atomize sequence of 30 seconds at 125°C, 30 seconds at 400°C, and 10 seconds at 2400°C. Prior to injection of 20 μL aliquots into the electrothermal atomizer, equal volumes of sample or standard are mixed with a palladium chloride matrix modifier [0.25 g of $PdCl_2$ dissolved in 10 mL of (1+1) nitric acid followed by dilution to 1 liter with ASTM Type II water] Corrections for background are required for these measurements.

7.27 PROCEDURES FOR TIN

Section 6.36 contains additional information for the determination of tin.

7.27.1 Determination of Tin; US EPA Method 7870

US EPA Method 7870 describes the determination of tin by atomic absorption spectrometry in the rich nitrous oxide–acetylene flame at 286.3 nm. Background correction is not necessary.

In a single-laboratory application of US EPA Method 7870 to the determination of tin in wastewater samples, the results were 4 ± 0.25, 20 ± 0.5, and 60 ± 0.5 ppm with recoveries of 95% or more.

7.28 PROCEDURES FOR THORIUM AND URANIUM

The atomic absorption resonance lines for thorium and uranium are at 324.6 and 351.5 nm, respectively. Measurements are usually made in a rich nitrous oxide–acetylene flame. Sensitivity for thorium is poor: 1% absorbance per 500 μg of Th per milliliter. Sensitivity for uranium is tenfold better.[172] However, ionization interferences of uranium must be suppressed by the addition of potassium or cesium chloride to both samples and standards.[173] Electrothermal atomization of uranium is inhibited by carbide formation.[174] Inductively coupled plasma–mass spectrometry is better suited to the identification and quantification of thorium and uranium.

7.28.1 Determination of Thorium and Uranium; US DOE Method 210

US DOE Method 210 is applicable to the determination of isotopes of thorium and uranium (^{230}Th and ^{234}U) in soil samples by inductively coupled plasma–mass spectrometry.[175]

A 1 g specimen of homogenized soil sample is weighed into a platinum crucible, moistened with 1 mL of water, and treated with 3 mL of hydrofluoric acid. The contents of the crucible are heated on a hot plate until dry, cooled, treated with a second 3 mL portion of hydrofluoric acid, and again heated to dryness. A third addition of hydrofluoric acid is made, and the contents of the crucible are again heated on a hot plate until dry.

To complete the fusion, 1.5 g of lithium metaborate ($LiBO_2$) is added to the crucible, and its contents are heated in with a gas flame to melt the flux. The hot crucible is placed in a beaker containing 50 mL of 5% (v/v) nitric acid, and dissolution of the fusion product is completed on a hot plate shaker. Finally, 20 mL of nitric acid is added to the beaker, and the contents are filtered into a 100 mL volumetric flask.

Prior to identification and quantification by inductively coupled plasma–mass spectrometry, the analytes are preconcentrated by flow injection analysis (FIA) using TRU-Spec ion exchange resin (EIChrome Industries) and 0.1 M ammonium oxalate eluant.

The FIA system is interfaced with the inductively coupled plasma–mass spec-

TABLE 7.3 Results for FIA/ICP-MS Determinations of Thorium and Uranium

	Soil TRM-4		Soil NRM-4	
sample	^{230}Th	^{234}U	^{230}2Th	^{234}U
Reference value, pCi/g	44.0 ± 1.6	44.6 ± 1.2	13.5 ± 1.1	11.4 ± 0.6
Mean results, pCi/g	44.4 ± 1.6	44.5 ± 1.0	12.3 ± 0.8	11.5 ± 0.3
Mean recovery, %	101	99.8	91.1	101

trometer so that the effluent is aspirated into the plasma. The ^{230}Th and ^{234}U mass peaks must be corrected for ^{232}Th background.

Detection limits are 0.1 pCi/g for ^{230}Th and 0.2 pCi/g for ^{234}U. Some results obtained for reference soils are presented in Table 7.3.

7.29 PROCEDURES FOR VANADIUM

Additional information for the determination of vanadium is contained in Section 6.38.

7.29.1 Determination of Vanadium; US EPA Method 7910

US EPA Method 7910 describes the determination of vanadium by atomic absorption spectrometry in a rich nitrous oxide–acetylene flame at 318.4 nm. Background corrections are required. Interferences from cobalt, chromium, iron, titanium, and phosphate can be compensated for by adding 1000 ppm aluminum at a rate of 2 mL per 100 mL of sample or standard. The concentration of vanadium in the samples is quantified by direct comparison of the absorbances to those of vanadium standards.

US EPA Method 7910 has been applied to the determination of vanadium in samples of wastewater prepared by US EPA Method 3010. Single-laboratory results were 2 ± 0.1, 10 ± 0.1, and 50 ± 0.2 ppm, with recoveries of 100, 95, and 97% respectively.

7.29.2 Determination of Vanadium; US EPA Method 7911

US EPA Method 7911 describes the determination of vanadium by electrothermal atomization–atomic absorption spectrometry. Atomic absorption measurements are made in nonpyrolytic graphite tubes at 318.4 nm with uninterrupted argon purge gas flow using a dry–ash–atomize sequence of 30 seconds at 125°C, 30 seconds at 1400°C, and 10 seconds at 2800°C. Background corrections are required for these measurements.

7.30 PROCEDURES FOR ZINC

Additional information for the determination of zinc is contained in section 6.39.

7.30.1 Determination of Zinc; US EPA Method 7950

US EPA Method 7950 describes the determination of zinc by atomic absorption spectrometry in a lean air–acetylene flame at 213.9 nm. Background correction is required. Addition of 1500 ppm strontium eliminates interferences from copper and phosphate ions.

8

METHODS FOR MONITORING TRACE ELEMENTS IN BIOLOGICAL TISSUES AND FLUIDS

8.1 ANIMAL TISSUES

Atomic spectrometry has become the method of choice both for monitoring exposures to toxic elements and for identifying systemic intoxication by many of them. In addition, atomic spectrometry has been applied to the assessment of nutritional status and to the diagnosis of disease. Blood and urine are tissues and fluids frequently used for such purposes. Subramanian[176] has reviewed the determination of two dozen trace elements in human blood by electrothermal atomization–atomic absorption spectrometry. Earlier, Delves[177] reviewed the applications of flame and nonflame atomic spectrometry to the determination of trace elements in blood and other tissues under normal and abnormal physiological conditions. Hair also has served as a biopsy specimen,[178] although the use of such tissues for these purposes is not widely recognized by the medical community.

8.1.1 NIOSH Method 8005 for Elements in Blood and Tissue

NIOSH Method 8005 is applicable to the analysis of blood or other tissues as an index for monitoring workplace exposures to the metals listed in Table 8.1.

Specimens of blood or other tissues from workers suspected of having been exposed and from controls known to be unexposed are collected and prepared by the procedures described under "Botanical and Biological Samples" in Section 3.1.2.2. Twenty metals are quantified by inductively coupled plasma–atomic emission spectrometry at the wavelengths listed in Table 8.1. Also listed in Table 8.1 are spike recoveries and accuracies for the metals determined by Method 8005.

TABLE 8.1 Recoveries of Metals from Blood

Metal	Wavelength (nm)	Spike (μg/100 mL)	Recovery	Accuracy (%)[a]
Antimony	217.58	10	106	15.6
Cadmium	226.5	10	120	22.2
Cobalt	231.2	10	81	60.2
Chromium	205.6	10	114	23.2
Copper	324.7	10	101	12.4
Iron	450.0	—[b]	—[b]	—[b]
Lanthanum	550.1	10	119	23.7
Lead	220.4	10	113	14.7
Lithium	670.8	10	113	15.2
Magnesium	279.6	10	114	27.5
Manganese	257.6	10	98	6.1
Molybdenum	281.6	10	126	32.1
Nickel	231.6	10	86	45.4
Platinum	203.7	10	92	35.4
Silver	328.3	10	115	16.6
Strontium	421.5	10	113	14.7
Thallium	190.9	10	97	20.0
Vanadium	310.2	10	131	33.2
Zinc	213.9	60	103	36.3
Zirconium	339.2	10	71	46.0

[a]NIOSH has defined "accuracy" as the degree of agreement between a measured value and the accepted reference value. The calculation of accuracy is made from the absolute mean bias of the method plus the overall precision at the 95% confidence level.

[b]Recovery not determined (blood iron concentration was above quantitation limit of spectrometer).

8.1.2 NIOSH Method 8310 for Metals in Urine

NIOSH Method 8310 is applicable to monitoring urine as an index to workplace exposures to the metals listed in Table 8.2.

Urine specimens from workers who may have been exposed and from controls known to be unexposed are collected and prepared by the procedures described under "Botanical and Biological Samples" in section 3.1.2.2. Sixteen metals are quantified by inductively coupled plasma–atomic emission spectrometry at the wavelengths listed in Table 8.2. Also listed in Table 8.2 are spike recoveries and accuracies for the metals determined by Method 8310.

8.1.3 Determination of Metals in Fish Tissue by Inductively Coupled Plasma–Atomic Emission Spectrometry; US EPA Method 200.11

US EPA Method 200.11 is applicable to the determination of aluminum, antimony, arsenic, beryllium, cadmium, calcium, chromium, copper, iron, lead, magnesium, nickel, phosphorus, potassium, selenium, sodium, thallium, and zinc in fresh fish tissues by inductively coupled plasma–atomic emission spectrometry.

TABLE 8.2 Recoveries of Metals from Urine

Metal	Wavelength (nm)	Spike (μg/100 mL)	Recovery	Accuracy (%)
Aluminum	308.2	20	100	17.2
Barium	455.4	0.4	80	41.6[a]
Cadmium	226.5	1.0	100	23.5
Chromium	205.6	1.0	100	15.3
Copper	324.8	10	100	8.2
Iron	259.9	40	100	11.6
Lead	220.4	10	100	7.8
Manganese	257.6	10	85	113[a]
Molybdenum	281.6	2.0	100	31.4[a]
Nickel	231.6	2.0	80	102[a]
Platinum	203.7	0.4	77	79.8[a]
Silver	328.3	2.0	100	23.5
Strontium	421.5	4.0	100	49.0[a]
Tin	190.0	2.0	100	41.2[a]
Titanium	334.9	2.0	86	45.5[a]
Zinc	213.9	200	100	17.4

[a]Does not meet the NIOSH criterion of ±25% accuracy (See Table 8.1).

Samples are collected by the procedures described in Section 2.2.6, and samples are prepared by the alkaline digestion method described in under "Botanical and Biological Samples" Section 3.1.2.2. When prepared samples are being aspirated into the inductively coupled plasma, care must be taken to avoid inclusion of floating and precipitated solids.

The recommended operating conditions are listed in Table 8.3, and emission lines recommended for measurement and background correction are listed in Table 8.4. Some results for NIST SRM 1566 (oyster tissue) obtained with US EPA Method 200.11 are presented in Table 8.5. The alkaline digestion used in US EPA Method 200.11 gave results comparable to those obtained with nitric acid–hydrogen peroxide digestion for most metals in quadruplicate specimens from two freshwater fish and two saltwater fish.

TABLE 8.3 ICP Operating Conditions for US EPA Method 200.11

Forward rf power	1100	W
Reflected rf power	< 5	W
Plasma viewing height	15	mm above coil
Coolant argon flow	19	L/Min
Carrier argon flow	630	mL/min
Plasma argon flow	300	mL/min
Sample uptake	1.2	mL/min

TABLE 8.4 Recommended Analytical and Background Wavelengths for US EPA Method 200.11

Analyte	Analytical Wavelength (nm)	Background Wavelength (nm)	Detection Limit (μg/g wet tissue)
Aluminum	308.215	308.276	0.3
Antimony	206.883	206.822	0.2
Arsenic	193.696	193.757	0.4
Beryllium	313.042	312.981	0.02
Calcium	315.887	315.826	
Cadmium	226.502	226.563	0.02
Chromium	205.552	205.522	0.05
Copper	324.754	324.693	0.05
Iron	259.940	260.001	
Lead	220.353	220.414	0.2
Magnesium	279.079	279.018	
Nickel	231.604	231.574	0.08
Phosphorus	214.914	214.944	
Potassium	766.491	766.430	
Selenium	196.026	195.965	0.6
Sodium	588.995	589.056	
Thallium	190.864	190.925	0.5
Zinc	213.856	213.886	0.07

TABLE 8.5 Accuracy and Precision of US EPA Method 200.11 Using NIST SRM 1566 Oyster Tissue

	Values (μg/g dry mass)	
Analyte	Certified	Experimental
Arsenic	13.4 ± 1.9	14.6 ± 0.2
Calcium	1500 ± 200	1560 ± 80
Cadmium	3.5 ± 0.4	3.39 ± 0.05
Chromium	0.69 ± 0.27	<0.02
Copper	63.0 ± 3.5	63.0 ± 1.5
Iron	195 ± 34	128 ± 16
Lead	0.48 ± 0.04	<0.8
Magnesium	1280 ± 90	1270 ± 30
Nickel	1.03 ± 0.19	1.28 ± 0.41
Phosphorus	(8100)	7360 ± 180
Potassium	9690 ± 50	9860 ± 50
Selenium	2.1 ± 0.5	<2.4
Sodium	5100 ± 300	4790 ± 110
Zinc	852 ± 14	832 ± 5

8.1.4 Determination of Trace Elements by Stabilized Temperature Graphite Furnace Atomic Absorption; US EPA Method 200.9

In addition to its applicability to the determination of trace elements in drinking water, groundwater, surface water, and wastewater, US EPA Method 200.9 is applicable to the determination of aluminum, antomony, arsenic, beryllium, cadmium, chromium, cobalt, copper, iron, lead, manganese, nickel, selenium, silver, thallium, tin, and zinc in biological tissues. Details on Method 200.9 are presented in Section 6.3.

8.1.5 NIOSH Method 8003 Lead in Blood and Urine

NIOSH Method 8003 quantifies lead in blood and urine for regulatory monitoring of occupational exposure and assessment of body burden.

Blood samples are collected in lead-free, heparinized, evacuated tubes such as the "blue-top" Vacutainer. The contents of the tubes are vortexed immediately after collection. Temperature for shipment and storage of the samples is 4°C; holding time at this temperature is 3 days.

Urine specimens are collected in clean, 125 mL, widemouthed polyethylene bottles and preserved by the addition of 0.2 mL of concentrated nitric acid. The preserved specimens are stable.

For the determination of lead, 2 mL aliquots of filtered urine or 2 g specimens of whole blood are transferred to culture tubes and treated with 0.2 mL of a solution made by dissolving 4 g of APDC and 2 mL of a nonionic detergent such as Triton X-100 in 40 mL of water followed by dilution with water to 200 mL. Distilled water blanks and positive and negative controls are treated similarly. The contents of the tubes are vortexed for 10 seconds, treated with 0.2 mL of water-saturated MIBK, and vortexed again for 2 minutes. The tubes are centrifuged for 10 minutes at 2000 rpm, and the lead contents of the upper MIBK phase are determined from their atomic absorptions in the air–acetylene flame at 283.3 nm relative to standards in the same matrix.

Lead recoveries from 42 blood samples spiked at concentrations ranging from 25 to 200 μg/100 mL and from 25 urine samples spiked at concentrations ranging from 10 to 100 μg/100 mL was 100% ±5%. Normal blood lead is below 25 μg/100 mL. Blood lead values in excess of 40 μg/100 mL are indicative of exposure, and blood lead values in excess of 60 μg/100 mL require corrective actions.

8.1.6 Determination of Mercury in Tissues by Cold Vapor Atomic Absorption Spectrometry, US EPA Method 245.6

US EPA Method 245.6 is applicable to the determination of total mercury in biological tissues. Samples are collected by the procedures described in Section 2.2.6.

A 200–300 mg specimen of tissue is placed on the bottom of the reaction vessel from the cold vapor generator. In some apparatus a 300 mL BOD bottle serves for this purpose. Positive and negative controls should be initiated at this time. To the contents of the reaction vessel are added 4 mL of sulfuric acid and 1 mL of nitric acid. The contents of the reaction vessel are heated in a 58°C water bath until the tissue specimen has dissolved. The contents of the reaction vessel are cooled in an ice bath and cautiously treated with five 1 mL increments of 5% (m/v) potassium permangenate solution. To ensure the maintenance of oxidizing conditions, an additional 10 mL of the permanganate solution is added. Next, 8 mL of 5% (m/v) potassium peroxydisulfate solution is added to the contents of the vessel, and the vessel is either allowed to stand overnight at room temperature or incubated for 30 minutes in a 80°C water bath. The contents of the reaction vessel are treated with 15 mL more of the permanganate solution and 8 mL more of the peroxydisulfate solution and incubated for an additional 90 minutes in a 30°C water bath.

The contents of the reaction vessel are diluted to 100 mL with ASTM Type II water and treated with 6 mL of 12% (m/v) sodium chloride–12% (m/v) hydroxylamine sulfate (or chloride) solution. The contents of each reaction vessel are treated individually for generation of the cold vapor and the measurement of its atomic absorption.

The reaction vessel is connected to the cold vapor generator, and the headspace is purged for 30 seconds to remove chlorine and other interferants. Then 5 mL of 10% (m/v) tin(II) chloride or tin(II) sulfate solution in 0.25 M sulfuric acid is injected into the reaction vessel, and the mercury vapor is swept into the absorption cell of the atomic absorption spectrometer, where absorbance at 253.7 nm is electronically recorded. The mercury content of the tissue specimen is determined from the height of the absorbance peak relative to peak heights from the standards.

Accuracy and precision evaluations of mercury recoveries from fish tissue samples fortified with methylmercury chloride reported 112 and 93% for homogenizates containing total mercury at concentrations of 0.19 ± 0.02 and 0.74 ± 0.5 μg/g.

8.2 PLANT TISSUES

The determination of metallic pollutants in vegetation and of metallic toxicants and mineral nutrients in food crops has been accomplished with atomic spectrometry.

8.2.1 AOAC Method 975.03 for Calcium, Copper, Iron, Magnesium, Manganese, Potassium, and Zinc in Plants[179]

Method 975.03 uses atomic absorption spectrometry to measure the concentrations of calcium, copper, iron, magnesium, manganese, potassium, and zinc in an acidic solution resulting from the dry or wet ashing of botanical tissues.

Adhering soil particles are removed from the botanical samples prior to drying and grinding, but excessive washing must be avoided to minimize leaching of water-soluble minerals from the plant tissues.

8.2.1.1 Dry Ashing Option A 1 g specimen of dried and pulverized plant material is weighed into a porcelain crucible and ashed at 500°C for 2 hours. The contents of the crucible are cooled to room temperature and moistened with 10 drops of water. Upon addition to the crucible of 3 or 4 mL of (1+1) nitric acid, nitric acid is evaporated at 100°C on a hot plate. The crucible is returned to the furnace for an additional hour of ashing at 500°C. The contents of the crucible are allowed to cool to room temperature, dissolved in 10 mL of (1+1) hydrochloric acid, and quantitatively transferred to a 50 mL volumetric flask.

8.2.1.2 Wet Ashing Option A 1 g specimen of dried and pulverized plant material is weighed into a 150 mL beaker and treated with 10 mL of concentrated nitric acid. After the content of the beaker have been thoroughly soaked with nitric acid, they are treated with 3 mL of perchloric acid and carefully heated until frothing ceases. Heating is continued and incremental additions of nitric acid are made as needed until a clear solution is obtained. The contents of the beaker are evaporated until white fumes of dichlorine heptoxide are evolved. The contents of the beaker are allowed to cool to room temperature, treated with 10 mL of (1+1) hydrochloric acid, and quantitatively transferred to a 50 mL volumetric flask.

8.2.1.3 Atomic Absorption Spectrometry The concentrations of calcium, copper, iron, magnesium, manganese, potassium, and zinc in solution in a volumetric flask are determined by atomic absorption spectrometry relative to standards in (1+1) hydrochloric acid under the conditions described in the tabulation that follows. Lanthanum chloride releaser and cesium chloride ionization suppressor must be added to both sample and standard solutions for the determination, respectively, of calcium and magnesium and of potassium.

Element	Wavelength (nm)	Flame Conditions
Calcium	422.7	Rich air–acetylene
Calcium	422.7	Rich nitrous oxide–acetylene
Copper	324.7	Lean air–acetylene
Iron	248.3	Rich air–acetylene
Magnesium	285.2	Rich air–acetylene
Manganese	279.5	Lean air–acetylene
Potassium	766.5	Lean air–acetylene
Zinc	213.8	Lean air–acetylene

8.2.2 AOAC Method 985.01 for Boron, Calcium, Copper, Magnesium, Manganese, Phosphorus, Potassium, and Zinc in Plants[179]

Method 985.01 is applicable to the determination of boron, calcium, copper, magnesium, manganese, phosphorus, potassium, and zinc in plant tissues. The sample is prepared by dry ashing, and the concentrations of the target elements are measured in the resulting solution by inductively coupled plasma–atomic emission spectrometry.

Adhering soil particles are removed from the botanical samples prior to drying and grinding, but excessive washing must be avoided to minimize leaching of water-soluble minerals from the plant tissues.

A 1 g specimen of dried and pulverized plant material is weighed into a porcelain crucible and ashed at 500°C for 2 hours. The contents of the crucible are cooled to room temperature and moistened with 10 drops of water. Upon addition to the crucible and of 3 or 4 mL of (1+1) nitric acid, nitric acid is evaporated at 100°C on a hot plate. The crucible is returned to the furnace for an additional hour of ashing at 500°C. The contents of the crucible are allowed to cool to room temperature, dissolved in 10 mL of (1+1) hydrochloric acid, quantitatively transferred to a 50 mL volumetric flask, and brought to volume with distilled water.

Quantification is by inductively- coupled plasma–atomic emission spectrometry as indicated at the following wavelengths (nm): boron, 249.6; calcium, 317.9; copper, 324.7; magnesium, 279.5; manganese, 257.6; potassium, 766.5; phosphorus, 214.9; zinc, 213.8.

9

METHODS FOR MONITORING TOXIC AND ESSENTIAL TRACE ELEMENTS IN FOODS

In addition to assessing potential hazards to human health from contaminants in air, water, and soil, atomic spectrometry has been applied to the identification and quantification of toxic elements in foods and food products. Atomic spectrometry has been applied also to the determination of nutritional minerals in food and food products.

The Nutritional Labeling and Education Act of 1990 (NLEA) mandated label information on nearly all processed foods. The Association of Official Analytical Chemists (AOAC) has prepared and evaluated several atomic spectrometric methods for monitoring toxic and essential trace elements in foods.[180,181] These manuals provide guidance to the methods for nutrient analysis.

Methods for the determination of trace elements in food by atomic spectrometry, like those for the determination of trace elements in air, water, and soil, are dynamic. They will continue to develop in response to changes in regulatory requirements.

9.1 AOAC METHODS 986.15 FOR ARSENIC, (CADMIUM, LEAD), SELENIUM, AND ZINC IN FOOD

Method 986.15 is applicable to the determination of arsenic, (cadmium, lead) selenium, and zinc. Samples are digested with nitric acid in closed vessels, and the concentrations of arsenic and selenium are measured by atomic absorption spectrometry of the gaseous hydrides. Flame atomic absorption spectrometry is employed for the quantification of zinc, and the concentrations of cadmium and lead are determined by anodic stripping voltammetry.

Samples (300 mg, dry basis) are weighed into the PTFE cups of steel-jacketed digestion vessels.[170] The contents of each cup, as well as empty cups serving as blanks, are treated with 5 mL of nitric acid. The vessels are closed and placed in a 150°C oven for 2 hours. After cooling to room temperature, the vessels are carefully opened in a fume hood, and the contents are transferred to 10 mL volumetric flasks. The contents of the flasks are brought to volume with water and divided into aliquots for arsenic, selenium, and zinc determinations.

9.1.1 Determination of Arsenic

For the determination of arsenic, an aliquot of the digested sample solution is pipetted into a 50 mL round, flat-bottomed flask and treated with 1 mL of a magnesium nitrate solution made by dissolving 3.75 g of MgO in 30 mL of water containing 10 mL of nitric acid and diluting the resulting solution to a final volume of 100 mL. The contents of the round, flat-bottomed flask are evaporated to dryness on a hot plate. The flask is then placed in a 450°C furnace to oxidize residual organic matter and to convert Mg $(NO_3)_2$ to MgO. After it has been cooled to room temperature, the residue in the flask is dissolved by the addition of 2 mL of 8 M hydrochloric acid and As(V) is reduced to As(III) by the addition of 0.1 mL of 20% potassium iodide solution. The contents of the round, flat-bottomed flask are transferred to the hydride generator, or the flask itself is connected to the device, whereupon arsine is generated by the addition of 2 mL of 4% sodium borohydride solution in 4% sodium hydroxide and atomized in the heated cell of the spectrometer, where its atomic absorption is measured at 193.7 nm. Blank corrections are applied, and arsenic is quantified from a calibration curve prepared with atomic absorption measurements on arsenic standards.

9.1.2 Determination of Selenium

For the determination of selenium, an aliquot of the digested sample solution is pipetted into a 50 mL round, flat-bottomed flask and treated with 1 mL of the magnesium nitrate solution described in Section 9.1.1. The contents of the flask are evaporated to dryness on a hot plate. The flask is then placed in a 450°C furnace to oxidize residual organic matter and convert $Mg(NO_3)_2$ to MgO. The residue in the flask is cooled to room temperature and is dissolved by the addition of 2 mL of 8 M hydrochloric acid. The mouth of the round, flat-bottomed flask is covered with a watch glass, and the contents are heated for 10 minutes on a steam bath. The contents are allowed to cool to room temperature and transferred to the hydride generator, or the flask is connected to this device. Hydrogen selenide is generated by the addition of 2 mL of 4% sodium borohydride solution in 4% sodium hydroxide and atomized in the heated cell of the spectrometer, where its atomic absorption is measured at 196.0 nm. Blank corrections are applied, and selenium is quantified from a calibration curve prepared with atomic absorption measurements on selenium standards.

9.1.3 Determination of Zinc

For the determination of zinc, an aliquot of the digested sample solution is pipetted into a 25 mL Erlenmeyer flask and treated with 0.1 mL of 1% (v/v) perchloric acid.

The contents of the flask are heated on a hot plate until white fumes of dichlorine heptoxide are observed and a clear, colorless solution is obtained. If charring occurs, incremental additions of 0.5 mL of nitric acid with intermittent reheatings are made until a clear, colorless solution is obtained. This is taken just to dryness, cooled, and redissolved in exactly 3 mL of 1% (v/v) perchloric acid. This solution is aspirated into the air–acetylene flame and its absorbance measured at 213.9 nm. Its zinc concentration is determined from a calibration curve prepared by measuring the absorbances of zinc standards at this wavelength.

9.2. AOAC METHOD 990.05 FOR COPPER, IRON, AND NICKEL IN EDIBLE FATS AND OILS

Method 990.05 is applicable to the determination of copper, iron, and nickel in edible fats and oils in the parts-per-million concentration range. Aliquots of oil samples, fat samples, or standards as organometallics in purified sunflower oil are injected into the electrothermal atomizer and their absorbances are measured under the conditions listed in Table 9.1.

The sunflower oil is purified by chromatography of a 1:3 dilution in "light petroleum" (bp, 40–60°C) on activated alumina. The purified oil is eluted with "light petroleum" and desolvated on a steam bath under nitrogen, and the last traces of solvent are removed under vacuum.

The niobium-coated graphite furnace tubes are prepared by alternately injecting 100 μL of 1000 ppm standard solution and performing the following cycle three times: dry, 60 seconds at 100°C and 5 seconds atomize at 2700°C; the drying–atomization cycles are continued until the absorbance at 302.1 nm is consistently low.

Samples and standards are maintained at 60°C to facilitate injection.

TABLE 9.1 Conditions for Absorbance Measurements According to Method 990.05

	Element		
Property	Copper	Iron	Nickel
Wave length, nm	324.7	302.1	232.0
Furnace Tube	Uncoated graphite	Niobium coated	Uncouated graphite
Ashing time (s)/ temperature (°C)	30/900	30/1200	30/1200
Argon flow, mL/min	300	300	300
Atomizing time (s)/ temperature (°C)	5/2700	5/2700	5/2700
Argon flow, mL/min	50	50	50

9.3 AOAC METHOD 991.25 FOR CALCIUM AND MAGNESIUM IN CHEESE

AOAC Method 991.25 is applicable to the determination of calcium and magnesium in various hard and soft cheeses. Samples are dried and ashed, the residues are dissolved in nitric acid, and the calcium and magnesium concentrations of the resulting solution are determined from their absorptions in the air–acetylene flame at 422.7 and 285.2 nm, respectively.

Samples of hard cheeses are homogenized in a food chopper. A food blender is used to homogenize samples of soft cheeses.

A 1 g specimen of homogenized cheese sample is weighed into a platinum or Victor crucible, dried for 1 hour in a 100°C oven, charred on a hot plate, and ashed overnight in a 525°C furnace. The ash is dissolved in 1 mL of nitric acid, and the resulting solution is transferred to a 250 mL volumetric flask. The contents of the flask are brought to volume with distilled water.

A 10 mL aliquot of the solution in the 250 mL volumetric flask is transferred to a 100 mL volumetric flask. Then 10 mL of lanthanum nitrate releaser solution (11.73 g of La_2O_3 dissolved in 25 mL of HNO_3 diluted to 1 liter) is added to the 100 mL volumetric flask, and the contents are brought to volume with distilled water. Mixed calcium and magnesium standards containing the lanthanum nitrate releaser are similarly prepared. Absorbances are measured for the sample and standard solutions in the air–acetylene flame. Calcium and magnesium are quantified relative to the standards at 422.7 and at 285.2 nm, respectively.

9.4 AOAC METHOD 984.27 FOR CALCIUM, COPPER, IRON, MAGNESIUM, MANGANESE, PHOSPHOROUS, POTASSIUM, SODIUM, AND ZINC IN INFANT FORMULA

In AOAC Method 984.27, samples of infant formula are digested with nitric acid–perchloric acid mixture, and the calcium, copper, iron, magnesium, manganese, phosphorus, potassium, sodium, and zinc in the resulting solution are quantified by inductively coupled plasma–atomic emission spectrometry.

To 1.5 g of powdered formula, 10 mL of concentrate liquid formula, or 15 mL of ready-to-feed liquid formula in a 100 mL Kjeldahl flask are added 30 mL of nitric acid–perchloric acid mixture (2+1) and three or four glass boiling beads. Reagent blanks are similarly prepared, and all the flasks are allowed to stand at room temperature overnight. Then, taking special care to avoid evaporation to dryness, the contents are cautiously heated until the nitric acid and water have been expelled. The digested samples and reagent blanks are transferred to 50 mL volumetric flasks and brought to volume with distilled water.

Relative to the emissions of standards in 20% perchloric acid, calcium (317.9 nm), copper (324.7 nm), iron (259.9 nm), magnesium (383.2 nm), manganese (257.6 nm), phosphorus (214.9 nm), potassium (766.5 nm), sodium (589.0 nm), and zinc (213.8 nm) in the resulting solution are quantified by inductively coupled plasma–atomic emission spectrometry at the wavelengths indicated.

9.5 AOAC METHOD 985.35 FOR CALCIUM, COPPER, IRON, MAGNESIUM, MANGANESE, POTASSIUM, SODIUM, AND ZINC IN READY-TO-FEED, MILK-BASED INFANT FORMULA

Residues from oven drying followed by dry ashing are dissolved in nitric acid prior to quantification of calcium, copper, iron, magnesium, manganese, potassium, sodium, and zinc by atomic absorption spectrometry under the conditions described in Table 9.2. Quantification of sodium is by flame (emission) photometry at 589.0 nm.

A 25 mL aliquot of sample is pipetted into a Victor crucible containing 5 g of filter pulp and dried in a 100°C oven. Blanks containing only filter pulp are also prepared. The dried residues in the crucibles are charred on a hot plate until smoking ceases and then ashed in a 525°C furnace until their white color indicates that they are carbon free. The ashed residues in the crucibles are dissolved by heating with 5 mL of 1 M nitric acid, and the resulting solutions are transferred to 50 mL volumetric flasks. To assure complete dissolution and transfer, the crucibles are treated twice more with 5 mL portions of the nitric acid. These are added to the volumetric flasks, the contents of which are then brought to volume with 1 M nitric acid.

Lanthanum chloride releaser is added to aliquots taken from the 50 mL volumetric flask for the quantification of calcium and magnesium. Matrix matching of the calcium and magnesium standards is required.

Cesium chloride ionization suppressor is added to the aliquots and standards for the determination of potassium (and sodium).

9.6 DETERMINATION COPPER AND NICKEL IN TEA

This AOAC method is applicable to the determination of copper and nickel in both cut and "instant" tea. Large samples are digested with nitric and perchloric acids,

TABLE 9.2 Wavelength and Flame Conditions for Quantification of Milk-Based Infant Formula by Method 985.35

Element	Wavelength (nm)	Flame Condition
Calcium	422.7	Reducing air–acetylene
Copper	324.7	Oxidizing air–acetylene
Iron	248.2	Oxidizing air–acetylene
Magnesium	285.2	Oxidizing air–acetylene
Manganese	279.5	Oxidizing air–acetylene
Potassium	766.5	Oxidizing air–acetylene
Zinc	213.9	Oxidizing air–acetylene

and copper and nickel are quantified in the resulting solutions by flame atomization–atomic absorption spectrometry.

A 3 g sample of cut tea or a 6 g sample of "instant" tea is weighed into a 400 mL beaker and treated with 100 mL of nitric acid. The large volume of acid makes necessary a reagent blank. The beaker is covered and allowed to stand for 10 minutes before being placed on a hot plate. The contents of the beaker are heated until the volume is reduced to a few milliliters. Upon addition of 50 mL more of nitric acid (for cut teas, 10 mL of perchloric acid), heating is continued until a clear solution is obtained. The contents of the beaker are transferred to a 50 mL volumetric flask and brought to volume with water.

Any precipitated potassium perchlorate should be allowed to settle before copper and nickel are quantified by atomic absorption spectrometry. Standards should be "matrix–matched" to the samples as follows: for cut teas, copper and nickel standards should contain 180 μg calcium/mL, 100 μg magnesium/mL, and 40 μg aluminum/mL in 8% (v/v) perchloric acid. For instant teas, the standards should contain 700 μg magnesium/mL, 130 μg calcium/mL, 70 μg sodium/mL, and 7000 μg potassium/mL in 10% (v/v) nitric acid.

Copper is quantified by comparison of standard and sample atomic absorptions in the air–acetylene flame at 324.7 nm. For nickel, the comparisons are made from the absorptions at 232.0 nm.

9.7 AOAC METHOD 985.40 FOR COPPER IN LIVER

With AOAC Method 985.40, low temperature, nitric acid digestion is used to prepare samples for the quantification of copper by atomic absorption spectrometry.

Specimens of liver having a wet weight of 1 g, or 0.25 g specimens of NIST SRM 1577 (bovine liver) are weighed into PTFE screw-cap bottles, treated with 5 mL of concentrated nitric acid, and incubated overnight in a 60°C oven. The contents of the bottles are transferred to 25 mL volumetric flasks and brought to volume with distilled water. The copper concentrations of the solution in the flasks are determined by atomic absorption relative to copper standards in (1+4) nitric acid at 324.7 nm in a lean air–acetylene flame.

9.8 DETERMINATION OF CADMIUM

This AOAC method describes digestion with sulfuric acid, nitric acid, and hydrogen peroxide, followed by extraction with dithizone in chloroform to prepare food samples for the flame atomic absorption spectrometry of cadmium at 228.8 nm.

A 50 g sample is weighed into a 1.5 L beaker and treated with 25 mL of nitric acid. Several boiling chips are added, and the contents of the beaker are gently heated to initiate the reaction. Three additional 25 mL increments of nitric acid are added with intermittent heating. Heating is continued with periodic additions of water from a wash bottle to control excessive frothing. Visible fatty and cellulose materials may be removed by chilling the contents of the beaker in ice water and decanting the cold solution through a glass wool filter. After the filtrate has been treated with 20 mL of sulfuric acid and diluted to 300 mL with water, it is heated until the volume is reduced and charring begins. When charring has become extensive, 1 mL increments of 50% hydrogen peroxide are cautiously added until the solution becomes clear. The clear solution is heated until dense white fumes of SO_3 are evolved. A blank is prepared from 50 mL of water in a 1.5 L beaker treated with equivalent amounts of the reagents used to prepare the sample.

Next, 2 g amounts of citric acid are added to the digested sample and to the reagent blank, which are diluted to 25 mL with water and treated with mL of thymol blue. With intermittent cooling between additions of ammonia, the pH is adjusted to 8.8 as indicated by a color change from yellowish green to greenish blue. After pH adjustment, the sample and the blank are transferred to 250 mL separatory funnels and diluted to 150 mL.

The solutions in the separatory funnels are extracted with 5 mL portions of freshly prepared 0.2% (m/v) dithizone in chloroform until the dithizone solution shows no color change when shaken with the contents of the separatory funnel. The dithizone extracts are combined in a 125 mL separatory funnel and washed once with 50 mL of water. The wash water is extracted with one 5 mL portion of chloroform, and this chloroform wash is added to the separatory funnel containing the dithizone extracts. The contents of the separatory funnel containing the dithizone extracts are extracted with 50 mL of 0.2 N hydrochloric acid to strip the cadmium (and some other metals) from the organic phase into the aqueous phase. The aqueous phase is washed with 5 mL of chloroform, transferred to a beaker, and carefully evaporated to dryness.

The residues in the beaker are dissolved in 5.0 mL of 2 N hydrochloric acid and aspirated into a lean air–acetylene flame for the measurement of atomic absorption at 228.8 nm. Quantification of cadmium in the samples is by direct comparison to aqueous cadmium standards.

9.9 DETERMINATION OF LEAD

Food can become contaminated with lead from the environment in which it was raised or grown (e.g. sewage-sludge-amended soils and soils in the vicinity of smelters and incinerators) and/or from the environment in which it was packaged (e.g. lead-soldered can seams). The source is of little consequence to the consumer.

9.9.1 Lead in Evaporated Milk

Samples of evaporated (condensed) milk are prepared by dry ashing, and lead is preconcentrated by extraction of the APDC complex with butylacetate prior to quantification by flame atomic absorption spectrometry at 283.3 nm with this AOAC method.

After being dried overnight in a 120°C oven, 25 g samples are ashed without ignition in a furnace, the temperature of which is raised in 50 degree increments from 250°C to 500°C. The sample is maintained in the 500°C furnace overnight, after which it is cooled and inspected for complete destruction of the organic matrix. If the ash is not white, it is moistened with a small amount of water and 1 or 2 mL of nitric acid, dried on a hot plate, ignited in a 500°C furnace for 1 or 2 hours, cooled, and reinspected for complete destruction of the organic matrix.

The residue is treated with 5 mL of 1 M nitric acid, warmed on a steam bath to ensure complete dissolution of lead, and filtered into a 50 mL volumetric flask.

Duplicate 20 mL aliquots are transferred from the volumetric flask into 60 mL separatory funnels. Lead standards and reagent blanks are transferred to 60 mL separatory funnels and diluted to 20 mL. To each separatory funnel are added 4 mL of 10% (m/v) citric acid solution and 2 or 3 drops of 0.1% (m/v) sodium salt of bromocresol green solution. The pH is adjusted to 5.4 as indicated by a color change from yellow to light blue with dropwise additions of ammonia. After pH adjustment, the sample, blanks, and standards are treated with 4 mL of 2% (m/v) APDC solution and mixed. The sample, blank, and standard solutions are extracted with 5 mL of butylacetate. After the phases have separated, the lower (aqueous) phase is removed and discarded.

Lead is quantified by flame atomization–atomic absorption spectrometry of the butylacetate phase at 283.3 nm. The instrument is adjusted to zero while water-saturated butylacetate is aspirated into a lean air–acetylene flame optimized for the organic phase.

9.9.2 Lead in Fish

A 25 g sample is weighed into a crucible and dried for 2 hours in a 135–150°C oven. The crucible containing the dried sample is placed in a cold furnace and slowly brought to a temperature of 500°C. The sample is ashed overnight at 500°C. The ash is cooled, treated with 2 mL of nitric acid, and heated at a low temperature until dry. The ash is placed in a cold furnace and slowly brought to a temperature of 500°C. After 1 hour, the ash is inspected for complete ignition as indicated by a white color. Additional treatments with nitric acid followed by drying and ignition cycles may be needed. When the materials are cool, the lead is dissolved from the ash by heating first with 10 mL then twice with 5 mL of 1 M hydrochloric acid. The hydrochloric acid solutions are filtered into a 25 mL volumetric flask and brought to volume with 1 M hydrochloric acid.

Lead is quantified relative to standards in 1 M hydrochloric acid by atomic absorption spectrometry at 283.3 nm in a lean air–acetylene flame.

9.9.3 Lead in Food

A sample containing at least 3 μg of lead having a dry mass of 10 g or less is weighed into a 500 mL Kjeldahl flask and treated with 1 mL of strontium scavenger (2 g of $SrCl_2 \cdot 6H_2O$/100 mL). An empty flask serving as the reagent blank is also treated with the scavenger. For each gram of sample dry mass is added 15 mL of "ternary acid" (100 mL of water containing 20 mL of sulfuric acid, 100 mL of nitric acid, and 70 mL of perchloric acid). The contents of the flask are allowed to stand for 2 hours. Digestion is completed by heating the contents of the flask until dense white fumes of SO_3 are evolved.

The contents of the flask are diluted with 15 mL of water and transferred while still warm to a 50 mL centrifuge tube. The contents of the tube are cooled and centrifuged for 10 minutes at 350 × *g*. The supernate is discarded, and the precipitate is retained in the centrifuge tube. After 20 mL of water containing 1 mL of sulfuric acid has been added to the Kjeldahl flask, the contents of the flask are heated and transferred to the centrifuge tube containing the precipitate. The contents of the tube are again cooled and centrifuged. The supernate is discarded, the liquid is drained from the centrifuge tube by inversion on clean absorbent paper, and the precipitate is carefully retained in the centrifuge tube.

The contents of the centrifuge tube are resuspended in 25 mL of saturated ammonium carbonate solution and allowed to stand for 1 hour to transpose the precipitated sulfates to the corresponding carbonates. The contents of the tube are centrifuged, and the supernate is discarded. The precipitate is resuspended in a second 25 mL portion of saturated ammonium carbonate solution and allowed to stand for 1 hour to assure transposition of the precipitate. The supernate is again discarded, and the liquid is drained from the centrifuge tube by inversion on clean absorbent paper.

The precipitate is dissolved in 5 mL of 1 M nitric acid, and after effervescence of CO_2 has ceased, the lead concentration of the solution is quantified by atomic absorption spectrometry relative to standards at either 217.0 or 283.3 nm in a lean air–acetylene flame.

9.10 DETERMINATION OF MERCURY

Concerns for public health have resulted in the development of methods for the determination of mercury by several agencies. Part of the history of Minamata disease comprises the tragic consequences of localized mercury pollution coupled with heavy consumption of mercury-contaminated sea fish. Equally tragic are the consequences of consuming bread made from grain dressed with a mercurial fungicide by peasants in Iraq and Pakistan.

9.10.1 Mercury in Food

A 5 g sample is weighed into a digestion flask and treated with 25 mL of (1+1) sulfuric acid, 20 mL of (1+1) nitric acid, and 2 mL of 2% (m/v) sodium molybdate solution. The digestion flask is fitted with a water-cooled condenser, and the contents of the flask are heated gently for 1 hour, then cooled for 15 minutes. Next, 10 mL each of nitric acid and perchloric acid are added to the flask through the condenser, and the flow of cooling water is stopped. The contents of the flask are heated until dense white fumes of SO_3 are evolved. The contents of the flask are allowed to cool, and 10 mL of water is added to the flask through the condenser. The contents of the flask are heated to boiling for 10 minutes. Heating is stopped, and three 10 mL portions of water are added to the flask through the condenser. The contents of the flask are transferred to a 100 mL volumetric flask and brought to volume with water.

A 25 mL aliquot is transferred from the volumetric flask to the reaction vessel of the cold vapor generator and diluted to a total volume of 100 mL with sulfuric acid–nitric acid mixture (67 mL sulfuric acid and 58 mL nitric acid per liter). Purge gas flow through the reaction vessel and spectrophotometer absorption cell is begun. The wavelength is set at 253.3 nm, the spectrophotometer is adjusted to zero, and 20 mL of "reducing solution" [15 g sodium chloride, 15 g hydroxylamine sulfate, 25 g tin(II) chloride, and 50 mL sulfuric per 500 mL] is injected into the reaction vessel. The mercury vapor is swept from the reaction vessel into the absorption cell from measurement of its atomic absorption. The mass of mercury in the food sample is quantified by direct comparison to mercury standards treated in the same manner.

9.10.2 Mercury in Fish

The bioaccumulation of mercury in fish tissue has been a matter of concern for more than a quarter-century. Airey's reviews on the subject[182,183] provide much information about human exposure to mercury from this source.

9.10.2.1 AOAC Method A 5 g specimen of homogeneous, wet fish tissue is weighed into a digestion flask and treated with 10 mL of sulfuric acid, 20 mL of nitric acid, and 10–20 mg of vanadium pentoxide. The digestion flask is fitted with a water-cooled condenser, and the contents of the flask are heated gently for 5 minutes and strongly for 10 minutes. After cooling for 15 minutes, 15 mL of water and 2 drops of hydrogen peroxide are added to the flask through the condenser. The contents of the flask are allowed to cool, transferred to a 100 mL volumetric flask, and brought to volume with water.

As an alternative, a 1 g specimen of homogeneous, wet fish tissue is weighed into the PTFE cups of a steel-jacketed digestion vessel. The contents of the cups, as well as an empty cup serving as a blank, are treated with 5 mL of nitric acid. The vessels are closed and placed in a 150°C oven for 0.5–1 hour. After being cooled to room temperature, the vessels are carefully opened in a fume hood, and the contents are transferred to 250 mL volumetric flasks. The contents of the volumetric flasks are brought to volume with water.

A 25 mL aliquot is transferred from the volumetric flask to the reaction vessel of the cold vapor generator and diluted to a total volume of 100 mL with a mixture of 67 mL sulfuric acid and 58 mL nitric acid per liter. Purge gas flow through the reaction vessel and spectrophotometer absorption cell is begun. The wavelength is set at 253.3 nm, the spectrophotometer is adjusted to zero, and 20 mL of "reducing solution" [15 g of sodium chloride, 15 g of hydroxylamine sulfate, 25 g of tin(II) chloride, and 50 mL of sulfuric per 500 mL] are injected into the reaction vessel. The mercury vapor is swept from the reaction vessel into the absorption cell from measurement of its atomic absorption. The mass of mercury in the food sample is quantified by direct comparison to mercury standards treated in the same manner.

9.10.2.2 EPA Method US EPA Method 245.6, Determination of Mercury in Tissues by Cold Vapor Atomic Absorption Spectrometry, has been successfully applied to fish. This method is described in Section 8.1.6.

9.11 AOAC METHOD 969.32 FOR ZINC IN FOOD

With AOAC Method 969.32, zinc in acid solutions resulting from either dry or wet ashing of food samples is quantified by flame atomic absorption spectrometry.

9.11.1 Dry Ashing Option for Method 969.32

A representative sample containing 25–100 μg of zinc is weighed into a platinum crucible, charred under an infrared lamp, and ashed without ignition in a 525°C furnace until the whiteness of the residue indicates that it is carbon free. The ash in the crucible is treated with a few milliliters of (1+1) hydrochloric acid and 20 mL of water. The contents of the crucible are evaporated to near dryness on a steam bath, treated with 20 mL of 0.1 M hydrochloric acid, heated for an additional 5 minutes, and filtered into a 100 mL volumetric flask. The crucible is rinsed with additional 10 mL portions of 0.1 M hydrochloric acid, and the rinsings are added to the flask through the filter.

9.11.2 Wet Ashing Option for Method 969.32

A sample having a maximum mass of 10 g and estimated to contain 25–100 μg of zinc is weighed into a 500 mL Kjeldahl flask and treated with 5 mL of concentrated nitric acid. The contents of the flask are heated cautiously until vigorous reaction ceases, whereupon 2 mL of concentrated sulfuric acid is added. Heating is continued with the addition of 2 mL increments of nitric acid as needed to maintain oxidizing conditions until a clear solution is obtained. The contents of the flask are then heated until nitric acid has been volatilized, as indicated by the appearance of dense white fumes. The contents of the flask are cooled, diluted with 20 mL of distilled water, and filtered into a 100 mL volumetric flask. The Kjeldahl flask is rinsed with additional 10 mL portions of water, and the rinsings are added to the volumetric flask through the filter.

9.11.3 Quantification of Zinc

The zinc concentrate of the solution in the volumetric flask is determined by atomic absorption relative to zinc standards at 213.9 nm in a lean air–acetylene flame. The zinc standards should be prepared in either 0.1 M hydrochloric acid or (1+49) sulfuric acid, depending on whether dry or wet ashing was used to prepare the sample.

10

LABORATORY MANAGEMENT

While financial and personal management are important parts of the laboratory organization, this chapter emphasizes the technical aspects.

Laboratories engaged in regulatory compliance monitoring provide a service to either internal or external clients. The mission of a service laboratory is to promptly and efficiently provide reliable results at a fair cost. To accomplish this mission, laboratory management should be concerned with:

- The adequacy of the facilities
- The competency of the personnel
- The capabilities of the equipment
- The limitations of the methodology
- The quality of the results
- The efficiency of communications
- The reputation of the organization
- The profitability of the operation

These concerns are interrelated, and deficiencies in any one will impede successful completion of the mission.

10.1 PHYSICAL FACILITIES

The physical facilities should provide a safe environment conducive to high levels of professional productivity. Ventilation and lighting coupled with adequate utilities connections are among the factors associated with a safe and productive laboratory.

The US EPA[184] has recommended the allocation of 150–200 ft^2 of work area to each member of the laboratory staff. The agency suggests further that each work area contain approximately 15 linear feet of usable bench space. Additional recommendations stress the need for a reliable source of high purity water and for adequate exhaust hood capacity to contain noxious and toxic fumes and vapors. The laboratory should be equipped with decontamination and fire suppression facilities: eye wash stations, safety showers, fire extinguishers, and so on. Provisions should be made for the safe use and storage of compressed gas cylinders and chemicals. Special considerations should be made for the disposal of chemical wastes.

The laboratory director should possess as minimum credentials an earned baccalaureate in an appropriate scientific discipline and 5 years of professional experience in regulatory compliance monitoring. An advanced degree is desirable, along with specialized training in spectrometric and chromatographic techniques. For a laboratory supervisor, the minimum credentials should be an earned baccalaureate in an appropriate scientific discipline and 2 years of experience in spectrometric and/or chromatographic techniques. The credentials of the quality assurance/quality control supervisor should be an earned baccalaureate in an appropriate scientific discipline and 2 years of experience in quality assurance/quality control techniques. The minimum credentials for a laboratory or field technician should be a high school or technical school diploma and either 2 years of experience or 1 year of experience and 1 year of post secondary education in laboratory or field procedures, with at least 6 months of on-the-job training for his or her responsibilities under the direct guidance of a laboratory supervisor. Continuing education is necessary to maintain and improve the professional skills of the technical personnel. Short courses, workshops, and seminars for this purpose are sponsored by regulatory agencies, professional organizations, and educational institutions.

When properly used, the instrumentation described in Chapter 1 and the equipment described in Chapter 2 are capable of yielding reliable results. The spectrophotometric instrumentation, as well as conductivity and pH/pI meters, and balances, require a well-defined program of inspection, calibration, and maintenance. An equally well-defined inspection/calibration/maintenance program is required for sample collection and sample preparation apparatus. Some aspects of these programs are described in Chapter 4.

The methodologies for regulatory compliance monitoring have undergone many reviews, revisions, and replacements, and they will continue to be modified and modernized. These changes have been, are, and will be driven by the need for better detection limits, the potential for matrix interferences, and improvements resulting from technological advances. The matrix interferences make a major contribution to the limits of identification and quantification. For example, analyte recoveries from potable water samples are superior to those obtained from samples of sewage sludge. It is necessary to recognize the limitations of the methodology and to validate the laboratory results with a rigorous quality assurance/quality control program. Some aspects of this program are described in Chapter 4.

The mission of the laboratory is to promptly and efficiently provide reliable results at a fair cost. While good people working in good facilities on good equipment with good procedures are likely to produce good results, the quality of the results is determined by the efficiency with which the laboratory is managed. It is important that all members of the laboratory staff, from the person who prepares the sample container to the person who writes the report, have a commitment to quality. A bad result is just as likely to originate from a mislabeled bottle, a faulty calculation, or a typographical error as it is to arise from an analytical error. The quality assurance/quality control program will minimize the consequences of analytical errors, but continual vigilance is required to guard against clerical errors.

Most communications between members of the technical staff are verbal. A manager can reinforce instructions or confirm directions with statements such as "I want to understand the procedure; tell me how the calibration is made," or "To make certain my report is accurate, describe the sample collection procedure for me." When a deficient response is received, further discussion is needed. An acceptable reply should be acknowledged with encouragement for success with the procedure.

Written communications are mandatory for the transmission of protocols, for the submission of proposals, and for the reporting of results. Each field and laboratory procedure must be completely described in a written protocol, and the protocol must be read and understood by participants in the field and laboratory operations. While proposals are submitted as written statements of work or lists of deliverables, they often involve oral presentations accompanied by slides/view graphs and handout materials. Oral communications are also important during the in-house development of protocols and proposals. To avoid misunderstandings, written records of oral communications should be maintained. These can be formalized as minutes and distributed among the principals. Reports of results must be written even when they have been transmitted by other means. The format and content must conform to what was agreed upon in the statement of work. A report can be as simple as a computer-generated tabulation of experimental values, or it can be a multipage interpretation of the laboratory findings.

Word of mouth and referrals are important in maintaining the reputation of the organization. Also important are the credentials of the technical staff, as shown by certifications and licenses from professional organizations and state or federal agencies, activity in professional organizations, service on boards and commissions of public bodies, and so on. Laboratory accreditation by governmental and professional agencies is mandatory.

While laboratory work should be enjoyable at all levels, it is necessary to productively manage resources to ensure that expenses will be met and that the owners will receive a fair return on their investment. The regulatory climate is influenced by political whims and economic pressures. Staff reductions must accompany reduced workloads, and bonuses are appropriate recognitions of superior efforts. Careful resource allocation and personnel utilization are needed to maximize efficiency and productivity during periods of "feast and famine."

10.2 LABORATORY ORGANIZATION

Frequently, laboratory organization is by task or function, and the number of tasks or functions assigned to an individual often depends on the size of the organization. Among the common tasks/functions are:

- Development of plans and protocols
- Preparation of sample containers and glassware
- Preparation of reagents and standards
- Collection of samples
- Transportation and storage of samples
- Preparation of samples
- Quantification of analytes
- Validation of data
- Preparation of reports

These are followed by billing/collection and customer service.

The division of labor usually relegates to technicians the preparation of the sample containers, glassware, reagents, and standards; the collection, transportation, and storage of samples; the preparation of samples, and the quantification of analytes. The laboratory director and supervisor(s) are responsible for project planning, protocol development, data validation, and report preparation.

10.3 MARKETING LABORATORY SERVICES

The market for laboratory services is highly competitive. A successful marketing plan for laboratory services requires a firm technical base, a competitive price, an on-site/off-site capability, and visible promotion.

Clients do not purchase laboratory services willingly. They are forced to do so to comply with monitoring regulations. The regulatory agencies have established standards to be met by the client. The client, in turn, will contract for laboratory services to satisfy the regulatory monitoring requirements. The proposal from the contract laboratory must present convincing evidence of technical competency. While it is difficult to clearly demonstrate superiority over the competition, it is often helpful to describe how facilities, personnel, equipment, and protocols have met state and/or federal certification requirements.

The client knows there is an economy of scale. Promotional literature should clearly indicate discounts on multiple samples and repeat samples. The client also knows extra services justify additional costs. Sample collection fees, priority service fees, sales tax, and so on should be shown in the promotional literature. Municipal contracts, which often have gender-based and race-based set-asides, usually go to the lowest qualified bidder. In bidding such contracts, exemptions from sales tax should be considered. In slack periods, the opportunity to work at a small profit should be balanced against the prospect of not working at all.

When developing proposals for long-term monitoring programs at sites remote from the laboratory, the possibility for establishing a satellite facility should be considered. Using rented equipment in a temporary structure may be more economical than transporting personnel and materials daily between the monitoring site and the central laboratory. This is especially true when the central laboratory is already operating near capacity. In addition, the client may be impressed by the idea of having a monitoring facility at the site.

Promotion of the laboratory should involve marketing vehicles ranging from the Yellow Pages to personal contacts. Valuable leads can be found among the legal notices, since municipalities are obligated to advertise bidding on contracts and requests for proposals. Direct mailing of promotional literature is a must. It may be helpful to first inquire by telephone for the name of the person most likely to be interested in the material, and to follow up with a call to determine whether there is immediate or future interest. The follow-up call will establish a personal connection for future contacts. Paid advertising in trade journals initiates and reinforces visibility.

10.4 PLANNING AND COORDINATING LABORATORY PROJECTS

The key to successful regulatory compliance monitoring is planning and coordination. The client must be involved in the planning process. Many of O. I. Milner's[185] anecdotal case histories exemplify this point. Here is a particularly relevant example.

> The operation of a process unit depended on the oxidizing power of vanadate solution. The process people had become accustomed to referring to the active ingredient as "vanadium" and submitted control samples to the laboratory for the determination of vanadium. The laboratory, not having been involved in the planning of the control testing and not being informed on the object of the analysis, analyzed the samples by oxidation–reduction. The method, of course, does not distinguish between the valence forms of the element and total vanadium was reported, leading to the false conclusion that no reagent was being consumed. The analyst could easily have determined pentavalent vanadium by a simple reductimetric titration had the laboratory been involved in planning the control tests or been consulted beforehand.

Another important reason for planning is to balance workload. During periods of underload, Milner suggests making the following short-term assignments as justifiable alternatives to layoffs: recalibration and restandardization of instruments and methods, training and rotation of personnel, preventive maintenance on equipment, quality control measurements, investigation of new methods, and housekeeping. Among his recommendation for handling overload are overtime for regular employees, utilization of part-time and/or temporary staff, and subcontracting some work to another laboratory.

The development of a work plan is important both to the client and to the laboratory. This detailed description of the project identifies the work to be done in terms of who, what where, when, how, and why: Who will do it? What will be done? Where will it be done? When will it be done? How will it be done? Why will it be done? Francks et al.[186] have suggested the following elements for inclusion in the work plan:

- Introduction
- Literature review and background analysis
- Conceptual model
- Rationale
- Description of tasks
- Project management
- Quality control/quality assurance
- Data evaluation
- Data presentation

These experts stress the need for proper planning with the "Rule of the Six P's: Proper Planning Prevents Pitiful Poor Performance."

Even with the best work plan, the project will not succeed without the firm guidance of a project manager. The project manager directs and coordinates the work plan, manages the human resources, maintains and adjusts the schedule for completion of tasks, makes contingency plans when necessary, oversees the budget, and reports to the supervisor and to the client.

10.5 REPORTING LABORATORY RESULTS

The laboratory report is often a computer-generated tabulation of the experimental values. The report must be dispatched to the client in an intelligible form and in a timely manner. Sometimes parallel reports are sent to a regulatory agency. This can be accomplished with conventional, hard copy mail, or by more rapid electronic communications. Copies of reports should be maintained for future reference.

Interpretive reports are multipage documents explaining the significance of the laboratory findings associated with in-depth or long-term monitoring projects or programs. A single project may require a preliminary report describing the objectives and the methodologies, interim or progress reports are timed intervals (usually quarterly), and a final report with recommendations and conclusions. Submissions of written reports may be accompanied by oral presentations. Francks et al.[186] have cited the importance of the report with "Clients see only two things from a consultant—a report and a bill. The report had better be good." They suggest the following elements for a report.

Letter of transmittal	The length of this can vary from one page to several pages. The purpose is to introduce the report to the reader.
Title page	Includes the subject of the report, for whom the report was prepared, who prepared the report, and the date.
Table of contents	Lists the divisions and subdivisions of the report, the figures, the tables, and the appendices.
Summary	Gives the casual reader something shorter and easier to read than the entire report. Particularly useful for reports read by corporate executives, legal counsel, etc. The summary includes an introduction to the project, the results, conclusions, and recommendations.
Report body	This is the full and detailed discussion of the project activities, findings, and detailed recommendations. Includes: (1) introduction and background; (2) discussion of project activities; (3) results and conclusions; (4) recommendations.
Appendices	Information in the appendix relates indirectly to the report body or has no logical place in the body of the report. Some organizations place all figures and tables in appendices, though it is generally a good idea to keep figures and tables closely pertaining to the text near the text to which they refer.

10.6 COMMUNICATING WITH CLIENTS, REGULATORS, AND THE PUBLIC

The client, the regulator, and the public often have different interests in the development, execution, and outcome of regulatory compliance monitoring. In addressing the concerns of each sector, it is frequently necessary to say the same thing in different ways.

Communications with the client include pre- and postproposal development conferences, the meetings for developing the work plan, and the sessions at which the reports are presented. Specifications for regulatory compliance monitoring are written into the agreement or contract between the principals. Contracts specify the rules and procedures for doing the work. Both client and contractor should understand that change orders, deviations from what was agreed upon, are changed to the contract and involve corresponding financial adjustments. The financial adjustments may be in either direction. To avoid redundancy and duplication of effort, each party to the contract should identify a point of contact in the other's organization for transmission of information. Telephone queries can receive verbal replies, but these should be followed by memoranda beginning "Pursuant to your telephone call of . . ." to document the communication. Fax and e-mail are rapid, efficient means of communication between client and contractor.

Communications with regulators should not be adversarial. To deal effectively with these parties, it is necessary to understand how the regulations impact the client, which regulatory compliance monitoring requirements apply to the project, and which agency has jurisdiction. Your participation in meetings between the client and the regulator, even if only as an observer, is often beneficial to all parties. The regulatory agency can provide information on the what, where, how, and why aspects of regulatory compliance monitoring. The regulatory agency will not become a party to the contract between the client and the contractor unless, of course, the agency is the client. However, communications between the regulatory agency and the client may be appended to the contract [e.g., "Samples will be collected on the first working day of each calendar quarter from each of the four monitoring wells located at the site described herein as specified in the attached letter of (DATE) from (AGENCY) to (CLIENT).

Public understanding of regulatory compliance monitoring is highly variable. It is best to claim "professional privilege" and avoid direct communications with the public. However, when the contract requires the contractor to make public statements on regulatory compliance monitoring, information must be presented in an honest and thoughtful manner. There are many misconception of risk assessments and maximum contaminant levels. For example, it is better (and safer) to present a soil chromium concentration as "below regulatory limits" than to cite a value of, say, 37 μg/g. To stray beyond the area of professional expertise is dangerous; that is, the analytical chemist may know about the carcinogenicity of some chromium compounds, but the toxicological considerations may be beyond the individual's scope of competency. Some members of the public have strong emotions about environmental issues. As a representative of the client, there is an obligation to refrain from hindering the resolution of these issues.

10.7 ETHICS IN LABORATORY MANAGEMENT

Most professional organizations publicize their codes of professional ethics, and many professional certifications and licenses require adherence to standards of professional conduct.

Some of the factors considered by Francks et al.[186] in connection with possible misfeasance, nonfeasance, or malfeasance are:

- Conflict of interest
- Dishonesty
- Competency
- Confidentiality
- Plagiarism
- Referrals

Management should develop and distribute a code of ethics that describes the professional liabilities and responsibilities of the members of the technical staff.

Without the approval of both parties and the court, accepting work from more than one litigant in the same action is a conflict of interest. Similarly, a conflict of interest would result if a full-time member of the technical staff were to engage in part-time work at a competing laboratory.

Dishonesty has many manifestations, ranging from violations of election laws to theft of services. Contributions to political parties and to political action committees are permitted, but the amounts and conditions have legal limitations. Laboratories engaged in bid/no-bid contracts with municipalities must be aware of laws governing elections. To do a favor for a friend by doing an "off the record" analysis of swimming pool water is dishonest. It is equally dishonest to bill a client for work that was not done.

To engage in professional activity beyond one's level of competency is unethical. The example was given earlier of the analytical chemist who may know about the carcinogenicity of some chromium compounds but is not competent to render an expert opinion on the toxicological considerations. It is unethical to claim or imply an expertise for which credentials have not been established. Even acknowledged experts protect themselves with malpractice insurance because they know they don't know everything.

The openness of scientific inquiry does not apply to the work done in a contract laboratory. Most activities of such laboratories are confidential. Proposal preparations are confidential to prevent disclosure of information that could aid a competing laboratory in submitting a lower bid for a contract. For similar reasons, client lists and negotiated fees are confidential. Laboratory data are confidential. Premature release of information on soil arsenic levels found during preconstruction monitoring could jeopardize financing for the building of a subdivision; rumors of lead-based paint on walls could hinder the sale of an office building. There is, however, an obligation to report public health hazards to the appropriate government agency. This is best done by the client.

To submit as one's own work a report from the open literature is plagiarism. Not only is this ethically wrong, it can lead to litigation, or at least to a damaged reputation, when discovered. It is equally unethical to claim the work of a subordinate as one's own.

Sometimes referrals for services beyond the capabilities of the laboratory are necessary, and sometimes "finder's fees" may be paid for referrals. Both are appropriate, but unnecessary referrals for the sake of the fee are certainly unethical and probably illegal. Referrals can be made by providing the client with nothing more than a list of specialized laboratories, or the list can be given with personal recommendations, or a specific recommendation can be made. Regardless of how the recommendation is made, it should be made with a clear understanding of whatever connections exist or do not exist between the laboratories.

REFERENCES

1. Grosser, Z. A., Inorganic Methods Update, *Environ. Test. Anal.* **1995**, *4*(3), 38–45.
2. Riley, R. G., Mong, G. M., Sklarew, D. S., Fadeff, S. K., Thomas, B. L., McCulloch, M., Goheen, S. C., Carter, M. H., and Morton, J. S., Developing DOE Methods, *Environ. Test. Anal.* **1995**, *4*(5), 30–35.
3. Product literature, Model Z-8200 Series Polarized Zeeman Atomic Absorption Spectrophotometers, Hitachi Instruments, Inc., Danbury, CT, 1991.
4. Product literature, GBC 932/933 Compact Atomic Absorption Spectrometers, GBC Scientific Equipment, Arlington Heights, VA, 1993.
5. Product literature, SIMAA 6000 Atomic Absorption Spectrometer, Perkin-Elmer Corp., Norwalk, CT, 1994.
6. Product literature, Flow Injection Mercury System (FIMS), Perkin-Elmer Corp., Norwalk, CT, 1994.
7. Product literature, Mercury Analysis, Thermo Separation Products, Fremont, CA, 1993.
8. Product literature, AP 200 Automated Mercury Preparation, PS 200 Automated Mercury Analyzer, Leeman Labs, Inc., Lowell, MA, 1992 and 1991.
9. Zagatto, E. A., Krug, F. J., Bergamin, H., and Jorgensen, S. S., Applications in Agriculture and Environmental Analysis, in *Flow Injection Atomic Spectroscopy,* Burguera, J. L., Ed., Marcel Dekker, New York, 1989, pp. 225–258.
10. Product literature, FIAS-100 and FIAS-400 Flow Injection Systems, Perkin-Elmer Corp., Norwalk, CT, 1992.
11. Robinson, J. W., *Atomic Spectroscopy,* Marcel Dekker, New York, 1990, p. 104.
12. Veillon, C., Analytical Chemistry of Chromium, *Sci. Total Environ.* **1989**, *86*, 65–68.
13. Product literature, Hitachi Instruments, Inc., Danbury, CT, 19??.
14. Product literature, PS1000AT Direct Reading Sequential Axial ICP, Leeman Labs, Inc., Lowell, MA, 1994.
15. Product literature, Baird ICP 2070, Baird Corp., Bedford, MA, 1992.

16. Product literature, JY 24 Sequential ICP Spectrometer and JY 38S Sequential ICP Spectrometer, JY Emission Division, Instruments S.A., Inc., Edison, NJ, 1993.
17. Product literature, GBC ICP, GBC Scientific Equipment, Arlington Heights, VA, 1993.
18. Product literature, Optima 3000 XL ICP Optical Emission Spectrometer, Perkin-Elmer Corp., Norwalk, CT, 1994.
19. Product literature, Thermo Jarrell Ash Corporation, Franklin, MA, 1994.
20. Pilon, M. J., Denton, M. B., Schleicher, R. G., Moran, P. M., and Smith, S. B., Evaluation of a New Array Detector Atomic Emission Spectrometer for Inductively Coupled Atomic Emission Spectrometry, *Appl. Spectrom.* **1990**, *44*(10), 1613–1620.
21. Kolczynski, J. D., Radspinner, D. A., Pomeroy, R. S., Baker, M. E., Norris, J. A., Denton, M. B., Foster, R. W., Schleicher, R. G., Moran, P. M., and Pilon, M. J., Atomic Emission Spectrometry Using Charge Injection Device (CID) Detection, *Am. Lab.*, May, 1991.
22. Part 136, Guidelines Establishing Test Procedures for the Analysis of Pollutants, 40 CFR Ch. 1 (7-1-93 edition), Appendix C, Table 1. Analyte Concentration Equivalents (mg/L) Arising from Interferants at the 100 mg/L Level, p. 579.
23. Application Note 1007, Analysis of TCLP Samples Using the Leeman Labs PS1000AT Axially Viewed Direct Reading Sequential Inductively Coupled Plasma Atomic Emission Spectrometer, Leeman Labs, Inc., Lowell, MA 19??.
24. Product literature, Fisons, PLC, Winsford, Cheshire, 1995.
25. Olesik, J. W., Fundamental Research in ICP-OES and ICP-MS, *Anal. Chem.*, N & F, Aug. 1, 1996, A469–A474.
26. Elements of ICPMS, *Anal. Chem.*, **1996**, *68*(1), A46–A49.
27. Creed, J. T., Brockhoff, C. A., and Martin, T. D., US EPA Method 200.8, Determination of Trace Elements in Water and Wastes by Inductively Coupled Plasma–Mass Spectrometry, Revision 5.4, United States Environmental Protection Agency, Cincinnati, OH, 1994.
28. Schoenleber, J. R., and Morton, P. S., Eds., *Field Sampling Procedures Manual,* New Jersey Department of Environmental Protection and Energy, Trenton, May 1992, pp. 79–81.
29. Designation D 1357, Standard Practice for Planning the Sampling of Ambient Atmospheres, *1993 Annual Book of ASTM Standards,* Vol. 11.03, American Society for Testing and Materials, Philadelphia, 1993, pp. 9–12.
30. Part 141, National Primary Drinking Water Regulations, 40 CFR Ch. 1 (7-1-93 edition) Section 141.23 Inorganic Chemical Sampling and Analytical Requirements, p. 608.
31. US EPA Method 1669, Sampling Ambient Water for Determination of Trace Metals at EPA Water Quality Criteria Levels, United States Environmental Protection Agency, Washington, DC, Draft, October 1994.
32. Keith, L. H., Principles of Environmental Sampling, *Environ. Sci. Technol.* **1990**, *24,* 610–617.
33. Rump, H. H. and Krist, H., *Laboratory Manual for the Examination of Water, Waste Water and Soil,* VCH, GmBH, Weinheim, 1988, p. 64.
34. Crépin, J., and Johnson, R. L., Chapter 2, Soil Sampling for Environmental Assessment, in *Soil Sampling and Methods of Analysis,* Carter, M. R., Ed., Lewis Publishers, Boca Raton, FL, 1993, p. 9.
35. RCRA Quality Assurance Project Plan Guidance, State of New York Department of Environmental Conservation, Albany, 1991.

36. Rieman, W., Neuss, J. D., and Naiman, B., *Quantitative Analysis,* McGraw-Hill, New York, 1951, pp. 13–14.
37. Kolthoff, I. M., Sandell, E. B., Meehan, E. J., and Bruckenstein, S., *Quantitative Chemical Analysis,* Macmillan, Toronto, 1969, pp. 516–518.
38. Robinson, J. W., *Undergraduate Instrumental Analysis,* 5th ed. Marcel Dekker, New York, 1995, pp. 30–31.
39. *Sampling Procedures and Protocols for the National Sewage Sludge Survey,* United States Environmental Protection Agency, Washington, DC, August, 1988.
40. *Test Methods for Evaluating Solid Wastes: Chemical/Physical Methods.* United States Environmental Protection Agency, Washington, DC, 1982.
41. Semu, E., Singh, B. R., and Selmer-Olsen, A. R., Mercury Pollution of Effluent, Air, and Soil Near a Battery Factory in Tanzania, *Water, Air Soil Pollut.,* **1986,** *27,* 141–146.
42. Mamuro, T., Matsuda, Y., Mizohata, A., and Matsunami, T., Activation Analysis of Airborne Dust, *Annu. Report/Radiat. Center of Osaka Prefecture,* **1971,** *12,* 1–9.
43. Katz, M., Ed., *Methods of Air Sampling and Analysis,* American Public Health Association, Washington, DC, 1981.
44. Designation D 4096–91, Standard Test Method for Particulate Matter in Atmospheres (High-Volume Sampler Method), *1993 Annual Book of ASTM Standards,* Vol. 11.03, American Society for Testing and Materials, Philadelphia, 1993.
45. Designation D 4536–91, Standard Test Method for High-Volume Sampling of Solid Particulate Matter and Determination of Particulate Emissions, *1993 Annual Book of ASTM Standards,* Vol. 11.03, American Society for Testing and Materials, Philadelphia, 1993.
46. Designation D 4185–90, Standard Practice for Measurement of Metals in Workplace Atmosphere by Atomic Absorption Spectrophotometry, *1993 Annual Book of ASTM Standards,* Vol. 11.03, American Society for Testing and Materials, Philadelphia, 1993.
47. Designation 5281–92, Standard Test Method for Collection and Analysis of Hexavalent Chromium in Ambient, Workplace, or Indoor Atmospheres, *1993 Annual Book of ASTM Standards,* Vol. 11.03, American Society for Testing and Materials, Philadelphia, 1993.
48. Method 6009, Mercury, in *NIOSH Manual of Analytical Methods,* 4th ed., United States Department of Health and Human Services, Cincinnati, OH, 1994.
49. Berg, T., Royset, O., and Steinnes, E., Blank Values of Trace Elements in Aerosol Filters Determined by ICP-MS, *Atmos. Environ.* **1993**, *27A*(15), 2435–2439.
50. *Manual of Methods for Chemical Analysis of Water and Wastes,* United States Environmental Protection Agency, Washington, DC, 1983.
51. Designation D 3694–89, Standard Practices for Preparation of Sample Containers and for Preservation of Organic Constituents, *1990 Annual Book of ASTM Standards*, American Society for Testing and Materials, Philadelphia, Vol. 11.02, 1990.
52. *Standard Methods for the Examination of Water and Wastewater*, 17th ed., American Public Health Association, Washington, DC, 1989.
53. Moody, J. R., and Lindstrom, R. M., Selection and Cleaning of Plastic Containers for Storage of Trace Element Samples, *Anal. Chem.* **1977**, *49*(14), 2264–2267.
54. Laxen, D. P., and Harrison, R. M., Cleaning Methods for Polythene Containers Prior to the Determination of Trace Metals in Freshwater Samples, *Anal. Chem.* **1981**, *53*(2), 345–350.

55. Kammin, W., Cull, S., Knox, R., Ross, M., McIntosch, M., and Thomson, D., Labware Cleaning Protocols for the Determination of Low-Level Metals by ICP-MS, *Am. Environ. Lab.* **1995,** *7*(9), 1, 6, 8.
56. *Handbook for Analytical Quality Control in Water and Wastewater Laboratories,* United States Environmental Protection Agency, Washington, DC, 1972.
57. US EPA Method 200.1, Determination of Acid-Soluble Metals, United States Environmental Protection Agency, Cincinnati, OH, 1991.
58. US EPA Method 200.2, Sample Preparation Procedure for Spectrochemical Determination of Total Recoverable Elements, United States Environmental Protection Agency, Cincinnati, OH, 1991.
59. US EPA Specification and Guidance for Contaminant-Free Sample Containers, Office of Solid Waste and Emergency Response, Directive 9240.0-05A, December 1992.
60. *Investigation of Matrix Interferences for AAS Metal Analysis of Sediments,* United States Environmental Protection Agency, Washington, DC, 1978.
61. Nackowski, S. B., Putnam, R. D., Robbins, D. A., Varner, M. O., White, L. D., and Nelson, K. W., Trace Metal Contamination of Evacuated Blood Collection Tubes, *Am. Ind. Hyg. Assoc., J.*, **1977,** *38,* 503–508.
62. Lecomite, R., Parades, P., Monaro, S., Barrette, M., Lamoureux, G., and Menard, H. A., Trace Element Contamination in Blood Collection Devices, *Int. J. Nuclear Med. Biol.* **1979,** *25*, 197.
63. Handy, R. W., Zinc Contamination in Vacutainer Tubes, *Clin. Chem.* **1979**, *25*, 197.
64. Versieck, J., Barbier, F., Cornelis, R., and Hoste, J., Sample Contamination as a Source of Error in Trace Element Analysis of Biological Samples, *Talanta,* **1982**, *29,* 973–984.
65. Koirtyohann, S. R., and Hopps, H. C., Sample Selection, Collection, Preservation and Storage for a Data Bank on Trace Elements in Human Tissues, *Fed. Proc.* **1981**, *40*(8), 2143–2148.
66. Katz, S. A., Collection and Preparation of Biological Tissues and Fluids for Trace Element Analysis, *Am. Clin. Prog. Rev.* April 1985, pp. 8–15.
67. Lee, R. E., and Duffield, F. V., Sources of Environmentally Important Metals in the Atmosphere, in *Ultratrace Metal Analysis in Biological Sciences and Environment*, Risby, T. H., Ed., American Chemical Society, Washington, DC, 1979, p. 148.
68. Ground Water Monitoring Program for Municipal Solid Waste Disposal Areas, 07-060-02, Nebraska Department of Environmental Quality, Lincoln, March 1994.
69. *Handbook: Stream Sampling for Waste Load Allocation Applications,* EPA/625/6-86/013, United States Environmental Protection Agency, Washington, DC, September 1986.
70. *Sampling Procedures and Protocols for the National Sewage Sludge Survey,* United States Environmental Protection Agency, Washington, DC, August 1988.
71. Hazardous Waste Monitoring System, *Fed. Regist.,* **1980**, *45*(98), pp. 33075–33127, May 19.
72. Kratochvil, B., and Taylor, J. K., *Anal. Chem.* **1981**, *53*, 924A–928A.
73. *Sampling for Toxic Substances*, Bendix Corp., Largo, FL, 1981.
74. Conway, R. A., and Malloy, B. C., Eds., *Hazardous Solid Waste Testing: First Conference*, ASTM STP 760, American Society for Testing and Materials, Philadelphia, 1981.
75. Conway, R. A. and Gulledge, W. P., Eds., *Hazardous and Industrial Solid Waste Testing:*

Second Symposium, ASTM STP 805, American Society for Testing and Materials, Philadelphia, 1983.

76. Jackson, L. P., Rohlik, A. R., and Conway, R. A., Eds., *Hazardous and Industrial Waste Management and Testing: Third Symposium,* ASTM STP 851, American Society for Testing and Materials, Philadelphia, 1984.
77. Petros, J. K., Lacy, W. J., and Conway, R. A., Eds., *Hazardous and Industrial Solid Waste Testing: Fourth Symposium,* ASTM STP 886, American Society for Testing and Materials, Philadelphia, 1985.
78. Lorenzen, D., Conway, R., Jackson, L. Hamza, A., Perket, C., and Lacy, W., Eds., *Hazardous and Industrial Solid Waste Testing and Disposal: Sixth Symposium*, ASTM STP 933, American Society for Testing and Materials, Philadelphia, 1986.
79. Eller, P. M., Ed., *NIOSH Manual of Analytical Methods,* 4th ed., United States Department of Health and Human Services, Public Health Service, Centers for Disease Control and Prevention, National Institute of Occupational Safety and Health, Cincinnati, OH 1994.
80. Wilkerson, C. L., Wehner, A. P., and Rancitelli, L. A., Leaching of Radioactive Nuclides for Neutron Activated Talc in Serum and in Dilute Hydrochloric Acid, *Food Cosmet. Toxicol.* **1977**, *15*, 589–593.
81. Versieck, J. M. J., and Speecke, A. B. H., Contamination Induced by Collection of Liver Biopsies and Human Blood. *Proceedings of the Symposium on Nuclear Activation Techniques in the Life Sciences,* Bled, 1972, International Atomic Energy Agency, Vienna, 1972.
82. Katz, S. A., and Salem, H., *The Biological and Environmental Chemistry of Chromium*, VCH Publishers, New York, 1994, pp. 156–157.
83. Masironi, R., and Parr, R. M., Collection and Trace Element Analysis of Post Mortem Human Samples: The WHO/IAEA Research Programme on Trace Elements in Cardiovascular Diseases, International Workshop on Biological Specimen Collection, Luxembourg, May 18–22, 1977.
84. Littell, B., Draft Work Plan for the 1994 (FY95) Regional Ambient Fish Tissue Monitoring Program, U.S. Environmental Protection Agency, Region VII, Kansas City, 1993.
85. Falandysz, J., Some Toxic and Trace Metals in Big Game Hunted in the Northern Part of Poland in 1987–1991, *Sci. Total Environ.* **1994**, *141*, 59–73.
86. Thompson, R. J., Collection and Analysis of Airborne Metallic Elements, in *Ultratrace Metal Analysis in Biological Sciences and Environment,* Risby, T. H., Ed., American Chemical Society, Washington, DC, 1979, p. 59.
87. Ehman, D. L., Anselmo, V. C., and Jenks, J. M., Determination of Low Levels of Airborne Chromium (VI) by Anion Exchange Treatment and Inductively Coupled Plasma Spectroscopy, *Spectroscopy*, **1987**, *3*, 32–35.
88. Arar, E. J., Long, S. E., Martin, T. D., and Gold, S., Determination of Hexavalent Chromium in Sludge Incinerator Emissions Using Ion Exchange Chromatography and Inductively Plasma Mass Spectrometry, *Sci. Total Environ.* **1992**, *26*, 1944–1950.
89. Preparation of Guidelines Establishing Test Procedures for the Analysis of Pollutants Under the Clean Water Act: Technical Amendments; Final Rule, *Fed. Regist.* (40 CFR Part 136), January 31, 1994, pp. 4504–4515.
90. Subramanian, K. S., Chakrabarti, C. L., Sueiras, J. E., and Maines, I. S., Preservation of Some Trace Metals in Samples of Natural Waters, *Anal. Chem.* **1978**, *50*, 444–448.

91. Truitt, R. E., and Weber, J. H., Trace Metal Ion Losses at pH 5 and 7, *Anal. Chem.* **1979**, *51*, 2057–2059.
92. Hoyle, W. C., and Atkinson, A., Retardation of Surface Absorption of Trace Metals by Competitive Complexation, *Appl. Spectro.* **1979**, *33*(1), 37–40.
93. Das, H. A., Faanhof, A., Gouman, J. M., and Ooms, P. C. A., Quantitative Aspects of the Absorption of Metal Ions from Aqueous Solutions to Container Surfaces, *J. Radioanal. Chem.* **1980**, *59*(1), 55–62.
94. Kingston, H. M., and Jassie, L. B., Microwave Energy for Acid Decomposition at Elevated Temperatures and Pressures Using Biological and Botanical Samples, *Anal. Chem.* **1986**, *58*, 2534–2541.
95. Product literature, Digestion Systems 6/20, Tecator AB, Höganäs, Sweden, 1988.
96. Product literature, 705 UV Digester, Metrohm Ltd., Herisau, Switzerland, 1991.
97. Smith, R. K., *Handbook of Environmental Analysis*, Genium Publishing, Schenectady, NY, 1993, p. 159.
98. D 3974 81, Standard Practice for Extraction of Trace Elements from Sediments, *1990 Annual Book of ASTM Standards*, Vol. 11.02, American Society for Testing and Materials, Philadelphia, 1990, p. 563.
99. D 4698–87, Standard Practice for Total Digestion of Sediment Samples for Chemical Analysis of Various Metals, *1990 Annual Book of ASTM Standards*, Vol. 11.02, American Society for Standards and Materials, Philadelphia, 1990, p. 636.
100. US DOE Methods for Evaluating Environmental and Waste Management Samples, MM210, ICP-MS of 99Tc, 230Th, and 234U Using Flow Injection Preconcentration, United States Department of Energy, Washington, DC, October 1994.
101. See Elements by ICP, NIOSH 7300, in Ref. 79.
102. Pundyn, A. M., and Smith, R. G., Application of Inductively Coupled Plasma Atomic Emission Spectrometry to Occupational Health, *J. Anal. Atomic Spectrom,* **1990**, *5*, 523–529.
103. Clark, T. A., Methods for the Determination of Metals in Environmental Samples, EPA/600/4-91/010, United States Environmental Protection Agency, Washington, DC, June 1991.
104. US EPA Method 200.8, Determination of Trace Elements in Water and Wastes by Inductively Coupled Plasma–Mass Spectrometry, Revision 5.4, EMMC Version, Environmental Monitoring and System Laboratory, United States Environmental Protection Agency, Cincinnati, OH, May 1994.
105. *Superfund Analytical Methods for Low Concentration Water for Inorganics Analysis*, United States Environmental Protection Agency, Alexandria, VA, October 1991.
106. US EPA Contract Laboratories Program Statement of Work for Inorganics Analysis, Multi-Media, Multi-Concentration, Document Number ILM01.0, IFB No. D000461R1/D000462E1.
107. D 5198-92, Standard Practice for Nitric Acid Digestion of Solid Waste, Compilation of Scopes of ASTM Standards Relating to Environmental Monitoring, *1993 Annual Book of ASTM Standards*, Vol. 11.02, American Society of Testing and Materials, Philadelphia, 1993.
108. Katz, S. A., and Jenniss, S. W., Determination of Some Macronutrients and Micronutrients and Some Toxic Elements in Sewage Sludges from Domestic and Industrial Influents Prior to Land Disposal: I. Development of Methods, in Lorenzen, D., Conway, R.,

Jackson, L., Hamza, A., Perket, C., and Lacy, W., Eds., *Hazardous and Industrial Waste Testing and Disposal*, Vol. 6, ASTM STP 933, American Society for Testing and Materials, Philadelphia, 1986, pp. 273–292.

109. Katz, S. A., Coordinated and Uncoordinated Evaluations of Sewage Sludge Reference Materials, *Fresenius J. Anal. Chem.* **1990**, *338*, 495–497.
110. US EPA Project Summary: An Interlaboratory Study of Inductively Coupled Plasma Atomic Emission Spectrometry Method 6010 and Digestion Method 3050, EPA/600/S4-87/032, January 1988.
111. Edgeli, K., US EPA Method Study 37, SW-846, Method 3050, Acid Digestion of Sediments, Sludges, and Soils, Contract EPA-68-03-3254, Bionetics Corp., Cincinnati, OH, 1989.
112. US EPA Method 3051, Microwave Assisted Digestion of Sediments, Sludges, Soils, and Oils, United States Environmental Protection Agency, Cincinnati, OH, September 1994.
113. Zehr, B. D., Van Kuren, J. P., and McMahon, H. M., Inorganic Microwave Digestions Incorporating Bases, *Anal. Chem.* **1994**, *66*, 2194–2196.
114. US EPA Method 200.3, Sample Preparation Procedure for Spectrochemical Determination of Total Recoverable Elements in Biological Tissues, United States Environmental Protection Agency Environmental Monitoring Systems Laboratory, Cincinnati, OH, April 1991.
115. US EPA Method 200.11, Determination of Metals in Fish Tissue by Inductively Coupled Plasma – Atomic Emission Spectrometry, United States Environmental Protection Agency, Environmental Monitoring Systems Laboratory, Cincinnati, OH, April 1991.
116. Murthy, L., Menden, E. E., Eller, P. M., and Petering, H. G., Atomic Absorption Determination of Zinc, Copper, Cadmium, and Lead in Tissues Solubilized by Aqueous Tetramethylammonium Hydroxide, *Anal. Biochem.* **1973**, *53*, 365–372.
117. Julshamn, K., and Andersen, K. J., A Study on the Digestion of Human Muscle Biopsies for Trace Metal Analysis Using Organic Tissue Solubilizer, *Anal. Biochem.* **1979**, *98*, 315–318.
118. NIOSH Method 8005, Elements in Blood or Tissue, in Ref. 79.
119. Binstock, D. A., Hardison, D. L., Grohse, P. M., and Gutknecht, W. F., Standard Operating Procedures for Lead in Paint by Hotplate- or Microwave-Based Acid Digestions and Atomic Absorption or Inductively Coupled Plasma Emission Spectrometry, RTI Report No. 91U-4699-100, Research Triangle Institute, Research Triangle Park, NC, 1991.
120. Designation D 5513-94, Standard Practice for Industrial Furnace Feedstreams for Trace Element Analysis, *Annual Book of ASTM Standards*, Vol. 11.01, American Society for Testing and Materials, Philadelphia, 1994.
121. Inductively Coupled Plasma Atomic Emission Analysis of Drinking Water, Appendix to Method 200.7, United States Environmental Protection Agency, Environmental Monitoring and Support Laboratory, Cincinnati, OH, March 1987.
122. US EPA Method 7195, Chromium, Hexavalent, Precipitation with Lead Sulfate, Test Methods for Evaluating Solid Waste, Physical/Chemical Methods, Technical Update, United States Environmental Protection Agency, Washington, DC, September 1986.
123. Hudnik, V., and Gomiscek, S., The Atomic Absorption Determination of Arsenic and

Selenium in Mineral Waters by Electrothermal Atomization, *Anal. Chim. Acta*, **1984**, *157*(1), 135–142.

124. US EPA Method 7197, Chromium, Hexavalent: Chelation–Extraction/Atomic Absorption, Test Methods for Evaluating Solid Wastes, Physical/Chemical Methods, United States Environmental Agency, Washington, DC, September 1986.
125. NIOSH Method 8003, Lead in Blood and Urine, in Ref. 79.
126. *Special Extraction Procedure, Methods of Chemical Analysis of Water and Wastes*, EPA-600-4-74-020, United States Environmental Protection Agency, Cincinnati, OH, 1979.
127. Richelmi, P. and Baldi, C., *Blood Levels of Hexavalent Chromium in Rats, Int. J. Environ. Anal. Chem.* **1984**, *17(3-4),* 181–186.
128. Smart, N., Lin, Y., and Wai, C. M., Supercritical Fluid Extraction of Metal Ions from Solid Samples, *Am. Environ. Lab.* **1996**, *8*(2), 38–42.
129. Rowan, J. T., and Heithmar, E. M., Pre-Concentration Method for Inductively Coupled Plasma–Mass Spectrometry, EPA Project Summary, EPA 600-54-98 043, 1990.
130. Long, S. E. and Martin, T. D., US EPA Method 200.10, Determination of Trace Elements in Marine Waters by On-Line Chelation Preconcentration and Inductively Coupled Plasma Mass Spectrometry, United States Environmental Protection Agency, Cincinnati, OH, 1992.
131. NIOSH Method 8310, Metals in Urine, in Ref. 79, 1994.
132. Hoffmann, P., and Lieser, K. H., Determination of Metals in Biological and Environmental Samples, *Sci. Total Environ.* **1987**, *64*, 1–12.
133. Fujiwara, K., Umezawa, Y., Numata, Y. I., Fuwa, K., and Fijiwara, S., Direct Atomization Atomic Spectrometry for Intact Biological Samples: Application to Trace Metal Analysis in Silkworm Eggs and Their Hatched Worms, *Anal. Biochem.* **1979**, *94*, 386–393.
134. Chakrabarti, C. L., Wan, C. C., and Li, W. C., Atomic Absorption Spectrometric Determination of Cd, Pb, Zn, Cu, Co and Fe in Oyster Tissue by Direct Atomic Absorption from the Solid State Using the Graphite Furnace Platform Technique, *Spectrochim. Acta*, **1980**, *35B*, 547–560.
135. Headridge, J. B., Determination of Trace Elements in Metals by Atomic Absorption Spectrometry with Introduction of Solid Samples into Furnaces: An Appraisal, *Spectrochim. Acta*, **1980**, *35B*, 785–793.
136. Mohamed, N. and Fry, R. C., Slurry Atomization Direct Atomic Spectrochemical Analysis of Animal Tissue, *Anal. Chem.* **1981**, *53*, 450–455.
137. Miller-Ihli, N. J., Slurry Sample Preparation from Simultaneous Multi-Element Graphite Furnace Atomic Absorption Spectrometry, *J. Anal. At. Spectrosc.* **1988**, *3*, 73–81.
138. Epstein, M. S., Carnrick, G. R., Slavin, W. and Miller-Ihli, N., Automated Slurry Sample Introduction for Analysis of River Sediment by Graphite Furnace Atomic Absorption Spectrometry, *Anal. Chem.* **1989**, *61*, 1414–1419.
139. Carnrick, G. R., Daley, G., and Fotinopoulos, A., Design and Use of a New Automated Ultrasonic Slurry Sampler from Graphite Furnace Atomic Absorption, *Atom. Spectrom.* **1989**, *10*(6), 170–174.
140. Katz, S. A., and Salem, H., *The Biological and Environmental Chemistry of Chromium*, VCH Publishers, New York, 1994, p. 83.

141. Kiilunem, M., Anttila, S., Aitio, A., Riihimäki, V., and Pääkkö, P., Nickel Concentrations in Tissues of Finnish Men Without Occupational Exposure, in *Nickel and Human Health*, Nieboer, E., and Nriagu, J. O., Eds., Wiley, New York, 1992, p. 119.

142. Designation D 1193-91, Standard Specification for Reagent Water, *1993 Annual Book of ASTM Standards*, Volume 11.03, American Society for Testing and Materials, Philadelphia, 1993.

143. Katz, S. A. and Jenniss, S. W., *Regulatory Compliance Monitoring by Atomic Absorption Spectroscopy*, Verlag Chemie International, Deerfield Beach, FL, 1983, pp. 76–77.

144. Part 141, National Primary Drinking Water Regulations Implementation, 40 CFR, Subpart B, 142.10 (b) (3) (i).

145. *New Jersey Register*, Monday, July 1, 1996, Regulations Governing the Certification of Laboratories and Environmental Measurements, CITE 28 N.J.R. 3333.

146. *Manual for the Certification of Laboratories Analyzing Drinking Water, Criteria and Procedures*, United States Environmental Protection Agency, Washington, DC, 1992.

147. Draft Constitution and Bylaws, National Environmental Laboratory Accreditation Conference, June 1, 1996.

148. *Guidance for Preparation of Combined Work/Quality Assurance Project Plans for Environmental Monitoring* (OWRS QA-1), United States Environmental Protection Agency, Washington, DC, May 1984.

149. *Manual of Methods for Chemical Analysis of Water and Wastes*, United States Environmental Protection Agency, Washington, DC, 1979.

150. Part 141, National Primary Drinking Water Regulations, 40 CFR, Subpart C, 141.23 (k) (5).

151. Part 136, Guidelines Establishing Test Procedures for the Analysis of Pollutants Under the Clean Water Act, 40 CFR, Table II.

152. *New Jersey Register*, Monday, July 1, 1996, Regulations Governing the Certification of Laboratories and Environmental Measurements, CITE 28 N.J.R. 3396-3398.

153. *New Jersey Register*, Monday, July 1, 1996, Regulations Governing the Certification of Laboratories and Environmental Measurements, CITE 28 N.J.R. 3367.

154. Anonymous, Survey of Reference Materials, Vol. 1, Biological and Environmental Reference Materials for Trace Elements, Nuclides and Micronutrients, International/Atomic Energy Agency, Vienna, 1995.

155. Anonymous, Survey of Reference Materials, Vol. 2, Environmentally Related Reference Materials for Trace Elements, Nuclides and Micronutrients, International Atomic Energy Agency, Vienna, 1996.

156. Becker, D., Christensen, R., Currie, L., Diamondstone, B., Eberhardt, K., Gills, T., Hertz, H., Klouda, G., Moody, J., Parris, R., Schaffer, R., Steel, E., Taylor, J., Watters, R., and Zeigler, R., The Use of NIST Standard Reference Materials for Decisions on Performance of Analytical Chemical Methods and Laboratories, NIST Special Publication 829, U. S. Government Printing Office, Washington, DC, 1993.

157. Taylor, J. K., Handbook for SRM Users, NIST Special Publication 260-100, U.S. Government Printing Office, Washington, DC, 1993.

158. Principles of Environmental Analysis, American Chemical Society, 1983 Reprint from *Anal. Chem.* December **1983**, *55*, 2210–2218.

159. US EPA 2185, Good Automated Laboratory Practices, Principles and Guidance to Reg-

ulations For Ensuring Data Integrity in Automated Laboratory Operations, United States Environmental Protection Agency, Washington, DC, August 10, 1995.

160. Nriagu, J. O., Ed., *Arsenic in the Environment, Part I: Cycling and Characterization* and *Arsenic in the Environment, Part II: Human Health and Ecosystem Effects*, Wiley, New York, 1994.

161. Sax, N. I., and Lewis, R. J., *Rapid Guide to Hazardous Chemicals in the Workplace,* Van Nostrand Reinhold, New York, 1986, p. 16.

162. Reference Method for the Determination of Lead in Suspended Particulate Matter Collected from Ambient Air, *Fed. Regist.* 52688, Dec. 1, 1976, as amended at 48 *Fed. Regist.* 2529, January 20, 1983, Appendix G to 40 CFR 50, National Primary and Secondary Ambient Air Quality Standards for Lead.

163. Katz, S. A., and Jenniss, S. W., *Regulatory Compliance Monitoring by Atomic Absorption Spectroscopy,* Verlag Chemie International, Deerfield Beach, FL, 1983.

164. Stevens, A. A., and Clark, T., Technical Notes on Drinking Water Methods, EPA-600/R-94-173, United States Environmental Protection Agency, Cincinnati, OH, October 1994.

165. *1995 Annual Book of ASTM Standards,* Section 11: Water and Environmental Technology, Vol. 11.01 Water (I) and Vol. 11.02 Water (II), American Society for Testing and Materials, Philadelphia, 1995.

166. 40 CFR 136, Guidelines for Establishing Test Methods for the Analysis of Pollutants Under the Clean Water Act: Technical Amendments, *Fed. Regist.*, 59, 20, 4504–4515, Jan. 31, 1994.

167. Rubio, R., Sahuquillo, A., Rauret, G., and Quevauviller, P., Determination of Chromium in Environmental and Biological Samples by Atomic Absorption Spectroscopy: A Review, *Int. J. Environ. Anal. Chem.* **1992**, *47*, 99–128.

168. Hazardous Waste Management System: Testing and Monitoring Activities, *Fed. Regist.* 60(9). Friday, Jan. 13, 1995, pp. 3089–3095.

169. US EPA Method 108, Determination of Arsenic Content in Ore Samples from Nonferrous Smelters, 40 CFR 61, Appendix B, 1993.

170. Katz, S. A. and Jenniss, S. W., *Regulatory Compliance Monitoring by Atomic Absorption Spectroscopy,* Verlag Chemie International, Deerfield Beach, FL, 1983, pp. 57–58.

171. US EPA Method 105, Determination of Mercury in Wastewater Treatment Plant Sewage Sludges, 40 CFR 61, 1993.

172. Robinson, J. A., *Undergraduate Instrumental Analysis,* 5th ed., Marcel Dekker, New York, 1995, pp. 380–382.

173. Dean, J. A., *Analytical Chemistry Handbook*, McGraw-Hill, New York, 1995, p. 7.25.

174. Welz, B., *Atomic Absorption Spectroscopy*, Verlag Chemie, Weinheim, 1976, p. 178.

175. *US DOE Methods for Evaluating Environmental and Waste Management Samples*, United States Department of Energy, OSTI, Oak Ridge, TN, 1995.

176. Subramanian, K. S., Determination of Trace Metals in Human Blood by Graphite Furnace Atomic Absorption Spectrometry, *Prog. Anal. Spectro.* **1986**, *9*, 234–237.

177. Delves, H. T., The Analysis of Biological and Clinical Materials, *Prog. Anal. Spectrom.*, **1981**, *4*, 1–48.

178. Chatt, A., and Katz, S. A., *Hair Analysis: Applications in the Biomedical and Environmental Sciences*, VCH Publishers, New York, 1988.

179. Sullivan, D. M., and Carpenter, D. E., 075.03—Metals in Plants, *Methods of Analysis for Nutritional Labeling,* AOAC International, Arlington, VA, 1993, pp. 152–155.

180. Boyer, K. W., Chapter 25; Metals and Other Elements at Trace Levels in Foods, in *Official Methods of Analysis of the Association of Official Analytical Chemists,* 14th ed., Williams S., Ed., Association of Official Analytical Chemists, Arlington, VA, 1984, pp. 444–476.

181. Sullivan, D. M., and Carpenter, D. E., *Methods of Analysis for Nutritional Labeling*, AOAC International, Arlington, VA, 1993.

182. Airey, D., Total Mercury Concentrations in Human Hair from 13 Countries in Relation to Fish Consumption and Location, *Sci. Total Environ.* **1983**, *31*, 157–180.

183. Airey, D., Mercury in Human Hair Due to Environment and Diet, *Environ. Health Prospect.* **1983**, *52*, 303–315.

184. Costle, D. M., *Manual for the Interim Certification of Laboratories Involved in Analyzing Public Drinking Water Supplies, Criteria and Procedures*, United States Environmental Protection Agency, Washington, DC, 1977.

185. Milner, O. I., *Successful Management of the Analytical Laboratory,* Lewis Publishers, Chelsea, MI, 1992, p. 96.

186. Francks, P. L., Testa, S. M., and Winegardner, D. L., *Principles of Technical Consulting and Project Management,* Lewis Publishers, Boca Raton, FL, 1992, pp. 51–64.

INDEX